GEORGE ADAMSKI

Letters to
Emma Martinelli

With an introduction and annotations
by Gerard Aartsen

George Adamski – Letters to Emma Martinelli
Introduced and annotated by Gerard Aartsen

First published April 2022.
Revised July 2023.

ISBN-13/EAN-13: 978-90-830336-2-4

Published on the occasion of the 70th anniversary of George Adamski's first contact in the desert 1952 - 2022

Typeset in Sans Source, Palatino, and Special Elite.

Cover design: Meryl Tihanyi.

“Unified understanding of life as we find it is the only thing that will awaken us to the everlasting presence of manifestations without a beginning or ending, and the eternal plan of the Divine Mind… Knowing all things, mysteries cease to be and divisions are replaced by oneness.”

George Adamski

Letter to Emma Martinelli, 8 May 1952

“To integrate, harmonize, and unify all things, and at the same time embrace all things in oneness and love, is the Telos *[goal] of all existence.”*

Ervin Laszlo Ph.D.

The Intelligence of the Cosmos – New Answers from the Frontiers of Science, 2017

George Adamski, circa 1950

Contents

Pioneers of Space
Cover and dust jacket of the original 1949 edition

INTRODUCTION

'Without full vision, man has divided the indivisible'

In 1949 George Adamski published his book *Pioneers of Space,* in which he describes a journey through space to the Moon, Mars and Venus. In his introductory statement Adamski admits that what we are about to read was "at present in the field of fiction", even though his descriptions were based on the scientific facts of his day, known law and common sense logic. However, he added his conviction that "the advance of science is so rapid that it will not be long before all of this will become a reality."[1]

The book was self-published in a print run of 350 copies and as Adamski was not yet a household name it was sold locally, bought by students or others who knew him from his time as a teacher with the Royal Order of Tibet in Los Angeles, Pasadena and Laguna Beach, as well as visitors to the Palomar Gardens Café along the Star Route in Valley Center, California where he and his group had been living and working since 1944.

By September 1951 the book was sold out and although Adamski was hoping to find a publisher for a second edition (see p.75), this never materialized. When he became world famous for his following books, *Flying Saucers Have Landed,* co-authored with Desmond Leslie in 1953, and *Inside the Space Ships* in 1955, *Pioneers of Space* fell into obscurity and remained largely unknown except to a few, until the first paperback

1 George Adamski (1949), *Pioneers of Space*, p.6.

reprint was published in 2008. It is currently available in three different paperback editions.

When it was found that some descriptions in this book bore a close similarity to passages in *Inside the Space Ships*, his acknowledgement that the experiences he described were "in the field of fiction" was quickly used to denounce his work wholesale as fiction, and merely so they can dismiss it as fantasy, many 'researchers' bluntly claim Adamski presented *Pioneers of Space* as "science fiction".

Emma P. Martinelli

After purchasing a copy of *Pioneers of Space* the month before, Emma Martinelli of San Francisco wrote the author in August 1950 about a number of questions she had, and to check if her understanding of certain concepts was correct. This was the start of a correspondence that lasted until at least May 1952, the date of Adamski's last known reply.

Emma Pearl Martinelli was born on December 29, 1901 in Sonoma, California into a family of judges who had settled in Marin County when her grandparents immigrated from Switzerland in 1848. She was the youngest sister of Marin County superior court judge Jordan L. Martinelli Sr, Ennio H. Martinelli, and Genevieve C. McNeil. Her father was Ennio Batista Martinelli, who served two terms as a state senator. Emma herself worked as a telephone operator for S&W Food Company in San Francisco from which she retired after 45 years.

According to her niece Elizabeth Cunningham, Emma never married – "her interest was in UFOs and in giving lectures on them. She was interviewed by the press on the radio and TV and was a guest speaker at a number of clubs. (...) As I remember my Aunt Emma, she had a hearty laugh and a sense of humor. We used to think she was slightly eccentric, but she had the courage to pursue her interest and belief in UFOs."[2]

Ms Martinelli was a long-standing member of the San Francisco Interplanetary Club, and in a letter to Riley Crabb, editor of the *Journal for Borderland Science Research,* in March 1962 she intimated that the care for one of her brothers and her demanding elderly mother left her frustrated that she couldn't spend more time on what she called her "world work". Given her paranormal experiences she said her "city brain in a country body" would have made her "a marvelous medium because I have my feet on the ground". That month the Journal had republished two of her letters about visits to Mount Shasta[3], which she also mentioned in a letter to Adamski (p.62), whom she knew from 1950 until his death in 1965.

By 1962 she had resigned from the SF Club, "due to not going along with Club policy", adding "there wasn't much more I could do for the club at this time (...) the theatrical part of the Saucers per se, is over. Now comes the work of spiritual reconstruction, and this is what most people don't go for – the gruelling task of making oneself over...".[4]

Emma died in Napa, California on June 26, 1996.

Projecting consciousness

In their exchange Ms Martinelli and George Adamski touch on a wide range of topics, from the mundane to the esoteric,

---

2 Tadashi Takeshima, 'George Adamski Archives' blog entry December 31, 2011. See: <georgeadamski.blog.fc2.com/blog-entry-2.html>.

3 'Probing the Mystery of Mt Shasta'. See: <borderlandsciences.org/journal/vol/18/n02/Martinelli_Mystery_of_Mt_Shasta.html>.

4 Letter to Riley Crabb, March 23, 1962.

and throughout Adamski shows his appreciation of his correspondent's understanding.

One of the major revelations in his letters to Emma Martinelli is Adamski's admission that his travels in *Pioneers of Space* should not be understood as physical but as out-of-body experiences, saying: "Yes, one may travel at will [to] any place in the universe without taking his physical body since the physical is not man, but rather the house of man." (p.63) And: "To you I can reveal this since your letter reveals much [understanding], while to others I keep silent about this." (p.38)

In the late 1950s twin brothers Ray and Rex Stanford, young flying saucer fans from Corpus Christi, Texas, visited Adamski at his Palomar residence. Reporting his experience twenty years later Ray said Adamski had told him: "Ray listen, I did not ever have to go out into space to know about the space ships and what was in 'em years ago... *Pioneers of Space* will tell you everything, just like *Inside the Space Ships*. All I did was project my consciousness to the beings out there and I could see them and know what was in their ships."[5]

Detractors have gratefully presented Mr Stanford's interpretation of Adamski's statement in 1978 as an admission of fraud and 'proof' of Adamski's 'deceit' ever since. Another fourteen years later, UFO researcher Jerome Clark, who first reported Mr Stanford's 'exposure' of Adamski, had duly simplified it to the point "that he [Adamski] had never been aboard a spaceship".[6] However, writing about the same

---

5 Jerome Clark, 'Startling new evidence in the Pascagoula and Adamski Abductions', *UFO Report*, Vol.6, No.2, August 1978. As quoted in Timothy Good (1998), *Alien Base – Earth's Encounters with Extraterrestrials*, p.150.

6 Jerome Clark (1992), *The Emergence of a Phenomenon. UFOs from the Beginning through 1959*, p.6.

experience William F. Hamilton, who drove the Stanford boys up to the Adamski residence in 1958, says: "Although Ray claims that Adamski virtually admitted that he had no need of contacts to describe what he had written in his books, I received no such impression from the man."[7]

Some may consider it a serious error of judgment when Adamski confided in a teenager, who subsequently 'exposed' him some twenty years later. However, it could also be argued that he recognized a certain sensitivity for spiritual matters in young Ray, who would soon begin his career as a trance medium, claiming to be the recipient of teachings from "Brothers" such as "Kuthumi", "Hilarion" and others from 1960 onward.[8] He also claimed he had been receiving "telepathic messages from the Space People" since 1954, when he was not yet 16 years of age.[9]

As will be discussed below, Adamski was very vocal about his criticism of psychics and their channelled 'messages', and the reader will have to decide if Mr Stanford's 'exposure' of Adamski twenty years after his visit was perhaps an attempt to get back at him. A quick glance at Stanford's conflicting claims over the years certainly does not bode well for the reliability of his statements.[10]

UK researcher Timothy Good doubts that his admission

---

7 William F. Hamilton III (1993), *Alien Magic*, p.14.

8 Since Koot Hoomi, Hilarion, and others were introduced as the names of some of the Masters of Wisdom in H.P. Blavatsky's Theosophical teachings in the late 19th century, many have claimed to have received their communications, but great discernment is needed to separate authentic teachings by the Masters from the wishful channelling of psychics or would-be scribes.

9 Ray and Rex Stanford (1958), *Look Up*, p.14.

10 See e.g. Douglas Johnson, 'Alien-lore Career Achievement Award goes to Ray Stanford', August 28, 2021, available at <alienexpanse.com/index.php?threads/alien-lore-career-achievement-award-goes-to-ray-stanford.5699/>.

necessarily means Adamski lied about all his actual, physical trips in space: "I think not. Assuming that is precisely what he said, he does not deny having made such trips; he seemed to be implying rather that it simply was not absolutely necessary for him."[11] Moreover, if his private concession of travelling in consciousness were an admission of fraud or misrepresentation, why would Adamski, in the same letter, write: "For even to me ... this book is like as if someone else wrote it. I can read it time and time again ... and marvel at the things that are in it. In other words, I have made the journey to get it." (p.39)

In his *Science of Life* course Adamski elaborates how consciousness may operate separately from the body that serves as its vehicle in the three-dimensional world of physical objects, and he describes an example of his ability to do so during his time as a metaphysics teacher in Laguna Beach. However, he says, "when my interest was taken up with flying saucers these experiences ceased"[12] – probably because they were no longer necessary when he was frequently invited on board the visitors' space craft. Also, as he explains how we may learn to direct our consciousness elsewhere, he explicitly distinguishes between his travels in consciousness and his "trips in space craft taken bodily".[13]

Based on the advancing insights of science into consciousness as the basis of reality we can't simply dismiss Adamski's statements as an attempt to counter the accusation of having fabricated the experiences in his books. To begin with, when he gave his explanation in his letters to Emma Martinelli

11 Timothy Good (1998), *Alien Base – Earth's Encounters with Extraterrestrials*, p.150.

12 Adamski (1964), *Science of Life* Study Course, Lesson Eleven.

13 Ibidem, Lesson Five.

in 1950 and 1951 he was not being accused of fabricating anything, and had merely stated in his introduction that he "used earthly measurements, scientific standards and relative logic and reason" in his description of the journey itself, "as an earthly foundation". (p.38)

More importantly though, in recent decades consciousness research by organisations such as the Institute of Noetic Sciences, founded by Apollo 14 astronaut Edgar Mitchell, the Scientific and Medical Network in the UK, the Campaign for Open Science, and the Laszlo Institute for New Paradigm Research founded by systems philosopher Ervin Laszlo, has led to the emergence of what has become known as post-materialist science. Proponents of this relatively new scientific paradigm hold that the institutional sciences are unable to prove that the brain is the creator of consciousness and is rather the receiver or, perhaps, the interface between consciousness and the three-dimensional world. They propose that consciousness and subjective experiences do not emerge from the physical brain or as the result of random processes in the evolution of matter, but exist *independent* of the brain. In fact, many hundreds of studies and reports of out-of-body or near-death experiences show that individual consciousness survives physical death.

Physicist Russell Targ, who was one of the pioneers of laser research in the 1960s, trained many people in 'remote viewing' – or projecting one's consciousness to seek impressions about a distant target – for various US intelligence agencies after he joined the Stanford Research Institute in 1972. Mr Targ calls remote viewing a natural ability that can be trained to describe and experience what is happening at a distant place. He says it is based on the fact "that we live in a non-local space-time", the quantum view of reality that says everything in the universe is connected – not just in theory, but in actual fact.[14]

In strikingly similar terms, but decades earlier, Adamski writes in his first letter to Emma Martinelli: "We are already everywhere. The only thing that governs location is the interest that the individual has [i.e. where he directs his attention; Ed.] ... If a man can lose location, there is only one other place he can find himself in – that is in the totality of the universe." (p.37)

In his writings and his talks Adamski often referred to the latest science as reported in newspapers and magazines, especially but not exclusively relating to astronomy and space travel, but I have not come across him mentioning the field of quantum physics anywhere. So was it prescience or personal ability and experience when Adamski spoke of the non-locality of consciousness – i.e. consciousness independent of a physical body – at a time when leading physicists were preoccupied with developing ever more destructive atom bombs?

Seen in the context of the post-materialist perspective of consciousness, we should at least consider the possibility that George Adamski's teachings and experiences were based on his abilities and higher knowledge, only known to his students, and in turn this does not preclude his later physical contacts and visits with extraterrestrials.

Observations in space

Critics claim that Adamski's observations as reported in *Pioneers of Space* have been shown to be a fantasy since

14 Russell Targ, 'Psychic abilities'. Talk for SUE Speaks, 2013. See: <www.youtube.com/watch?v=pVZ24r3y5_U>.

astronauts from Earth went to the Moon, but Desmond Leslie pointed out several examples of observations that Adamski made which were only later confirmed by astronauts.

During the first American space flight orbiting Earth in February 1962, Mercury 7 astronaut John Glenn reported from his Friendship capsule: "I'm in a big mass of thousands of very small particles that are brilliantly lit up like they're luminescent. They are bright yellowish-green. About the size and intensity of a firefly on a real dark night. I've never seen anything like it."[15]

According to astronauts Alan Shepard and Deke Slayton in their book *Moon Shot* (2011), the mystery of the 'fireflies' in space was solved on the next orbital flight when a swarm of 'fireflies' appeared after astronaut Scott Carpenter banged his hand on the inside wall of his Aurora 7 capsule and he concluded the 'fireflies' effect was caused by condensation on the outer wall when he tapped the inside, or when vapor was being vented from the spacecraft.

Remarkably, about his first trip "taken in the physical" – in February 1953 – Adamski writes: "... as I looked out I was amazed to see that the background of space is totally dark. Yet there were manifestations taking place all around us, as though billions upon billions of fireflies were flickering everywhere, moving in all directions..."[16]

Despite the similarities between *Inside the Space Ships* and *Pioneers of Space*, the 'fireflies in space' description is not found in the latter book. This supports my suggestion that his statement

---

15 Jay Barbree, 'A glowing mystery surrounds the first American in orbit', NBC News, 12 May 2011. See: <www.nbcnews.com/id/wbna42982294#.WEnkvOErJ0s>.

16 Adamski (1955), *Inside the Space Ships*, p.76.

in his Foreword: "Upon these facts, known law and common-sense logic we base what you are about to read", should be interpreted as saying that he chose to describe the journey itself in terms of a physical trip by rocket ships to conceal his actual mode of 'transport', i.e. projecting out his consciousness.

Elsewhere in *Inside the Space Ships* Adamski describes bands of radiation surrounding the Earth that were only later discovered by astronomers and named the 'Van Allen Belt'. He also reported the space craft he was travelling on entering a glowing patch of light. Similar reports later came from astronauts Walter Schirra and Gordon Cooper during their respective Mercury missions, as well as from Apollo 7 astronaut Walter Cunningham, whose observations were documented in the 1968 Condon Report.[17]

As Desmond Leslie asked in 1970, how did Adamski know about these things in 1953, unless he had actually seen them for himself? "Lucky guesses? *Adamski made too many 'lucky guesses' for comfort.*"[18] And in response to accusations that *Inside the Space Ships* was a reworking of *Pioneers of Space,* Timothy Good said: "... apart from those similarities (...), *Inside the Space Ships* is a very different book, full of much more richly detailed descriptions...".[19]

Some critics supported their allegation that Adamski fabricated his account in *Inside the Space Ships* by accusing him of taking his description of a repulsive force employed

17 Edward U. Condon (1968), *Scientific Study of Unidentified Flying Objects.* Funded by the US Air Force, this study is seen as having been instrumental in diminishing the academic interest in UFOs.

18 Desmond Leslie, 'Commentary on George Adamski', in *Flying Saucers Have Landed*, Revised & Enlarged Edition (1970), pp.242-43.

19 Good (1998), op cit, p.150.

by the visitors' space craft from Frank Scully's book *Behind the Flying Saucers* (1950). But, as Adamski pointed out in an article in 1951[20], he had already described such a repulsive force in *Pioneers of Space* in 1949.[21]

Paranormal communication and confusion

On various occasions in his letters to Ms Martinelli, as well as elsewhere, Adamski denounces mystics (i.e. psychics and mediums) who claim to 'channel' messages of deceased people or even cosmic beings. As he once wrote to a supporter: "There are indeed very few psychics on whose messages you can bank. There is no doubt they are hooking onto something, but surely it is of no value to the present situation. It is these things that are creating more confusion than stability in the minds of the people who are sincerely seeking the truth about present-day events."[22]

Later he clarified why communications coming from trance mediums are unreliable at best: "Two and a half billion people [the world population at the time; Ed.] are broadcasting thoughts of expectation as to what they think will occur. Most prophecies are nothing more than the influence of these thoughts. People with little understanding of the human mind receive these thoughts and believe they are messages from space people or a revelation from God."[23] As he once told his Swiss co-worker Lou Zinsstag: "Every mock spirit

---

20 Adamski, 'I Photographed Space Ships'. *FATE* magazine Vol.3, No.21, July 1951, p.72. See also page 56 in this volume.

21 Adamski (1949), op cit, pp. 74, 148, and 199.

22 Letter to Laura Mundo, January 4, 1955. In: *The Adamski Documents, Part One* (1980), p.7.

23 Adamski (1962), 'World Disturbances'. *Cosmic Science*, Vol. 1, No.1, p.4.

or evil impersonator could come and tell you that his name was 'Ashtar' or 'Aetherius' and that he lived in a space ship. I think that these entities are having a heyday leading astray the gullible mediums and their public."[24]

Critics find this difficult to square with his own claims to telepathic abilities and see it as evidence of his untrustworthiness – claiming 'paranormal' or 'extrasensory' capabilities for himself, while denying the reliability of others. The apparent contradiction, however, results when we fail to distinguish between trance mediumship and deliberate communication between two conscious minds, which involve very different types of non-local sources and entirely different techniques. The latter refers to communication between the consciousness of actual individuals – in or out of incarnation – while the former usually involves the passive reception by the medium of astral thoughtforms or a coagulate of someone's memories, hopes and fears, plus the medium's own conditioned emotions or projections. And this is what Adamski objects to and rejects as unreliable. (While more and more scientists are exploring the field of non-local consciousness as a fundamental aspect of reality, a critical classification of communications requires more discernment based on a deeper understanding of the composition of man and the various planes on which life manifests, such as can be found in the Ageless Wisdom teachings.[25])

With today's knowledge, it should not surprise us that Adamski didn't reject wholesale what is often called paranormal communication, despite his repeated criticism of psychics. In his article for *FATE* magazine in 1951 he refers to a massive

24 Lou Zinsstag & Timothy Good (1983), *George Adamski – The Untold Story*, p.55.

25 See also the Appendix on page 101.

sighting of a space ship during a meteor shower in October 1946, which was also reported in the press (see p.14). Here he describes going back inside after the sighting and turning on the radio: "We were all amazed to hear the announcer telling of an immense space ship which hundreds of people in San Diego and for miles up the coast had seen while they were watching the shower. So great was the interest of some of the people there that they had prevailed upon a well known spiritualist medium to attempt to project her mind out into space, follow the ship and find out who they were and why they had not landed on earth when they had been so close. (...)

"Later the medium's report was given over the radio. She was reported to have said that this ship had come from one of the other planets in our system; that its occupants had not landed on earth because they were not sure how they would be received by earthmen. The name of the planet was not given."[26]

Citing this episode at length, Adamski publicly hinted at what he privately shared with Emma Martinelli about projecting one's consciousness out of the body to observe things at a distance, as the means used for his experiences in *Pioneers of Space*. Seeing his private disclosure to Ms Martinelli – that his journey was an out-of-body experience – alongside his public reference to exactly such an approach described in a radio broadcast about a mass sighting of a space craft, without disqualifying it, provides a fuller understanding of Adamski's actual views, which is indispensible for a correct perception of his mission and has clearly been missing in most assessments, both favourable and scorning, of his work.

To be sure, Adamski himself was very well aware of

---

26 *FATE* magazine, op cit, pp.65-66.

'Space Ship' Seen By San Diegans Trying To Contact Earth

LOS ANGELES—The Los Angeles Daily News said in a story from San Diego, that many persons there believed a "space ship from another planet" was trying to contact earth during last week's showering of meteors from the Giacobini-Zinner comet.

At San Diego, authorities said they received no reports of such a phenomenon and, in fact, were frankly sceptical of it. Police said there were no calls about it at the time.

The Daily News said at least a dozen persons testified that on the night of Oct. 9 a "large and weird object," with something that looked like wings was in motion over the city.

SHAPED LIKE BULLET

"It was shaped like a bullet and it left a thin vapor trail behind it," the News quoted William L. Nabers, a San Diego county hospital attendant, as saying.

Meade Layne, editor of an occult publication called the "Round Robin," was reported as putting a medium to work on the supposed sighting. The medium claimed it was a "space ship named Careeta" from an unidentified planet, the News said. He added, the "ship" didn't land because "they're afraid of the reception they'll get."

San Diegoans See (?) Visitor

LOS ANGELES —(U.P.)— The Los Angeles Daily News said in a story from San Diego, Calif., Monday that many persons there believed a "space ship from another planet" had tried to contact earth during last week's showering of meteors from the Giacobini-Zinner comet.

At San Diego, authorities said they received no reports of such a phenomenon and were frankly skeptical of it. Police said they had received no calls at the time.

The Daily News said at least a dozen persons testified that on the night of Oct. 9 a "large and weird object," with something that looked like wings had appeared over the city.

"It was shaped like a bullet and it left a thin vapor trail behind it," the News quoted William L. Nabers, San Diego County Hospital attendant, as saying.

Meade Layne, editor of an occult publication called the "Round Robin," was reported as putting a medium to work on the supposed sighting. The medium claimed it was a "space ship named Careeta" from an unidentified planet, the News said. He added the "ship" didn't land because "they're afraid of the reception they'll get."

Reports about the mass sighting of a space ship over San Diego on October 9, 1946 appeared in the *Register-Guard* of Eugene, Oregon, October 14 (right), and the *Marine Corps Chevron* of San Diego, California, October 18, 1946 (above).

the distinction between astral psychism or mediumship – "communicating with spirits" – on the one hand, and mental telepathy with higher beings or sources of consciousness. In *Flying Saucers Farewell*, after giving definitions of terms like "psyche", "occultism", "theosophy" and "philosophy", he asserts: "A little thought on the subject will quickly show one that present-day psychism, compared with the real meaning as [in the definitions] given above, is nothing more than a perversion of the true science of the occult, metaphysics, etc."[27]

Finally, when Adamski claims his journey in *Pioneers of Space* took place as an out-of-body experience by projecting his consciousness afar, this builds directly on his teaching about consciousness in his first book, *The Invisible Ocean*. It also portends how he later taught his students to acquire the same ability in his *Science of Life* course, just like 'remote viewing' was researched and trained by members of various international intelligence agencies.

As pointed out elsewhere, in both these publications he referred to the universe as a sea of consciousness that we traverse on the ladder of evolution, and from which everything we see descends into objective reality. Very much the same view is now held by an increasing number of theoretical physicists, systems philosophers and consciousness researchers, based on the latest findings in quantum research.[28]

The nature of reality

Seeing how the confusion caused by psychics claiming to channel messages from their imaginary space friends

27 Adamski (1961), *Flying Saucers Farewell*, pp.111-12.

28 See Gerard Aartsen (2019), *The Sea of Consciousness*.

jeopardized his mission of presenting the reality of the space visitors as actual, albeit extraterrestrial human beings, Adamski emphasized whenever possible that his contacts were with physical beings from other planets who come here in physical space craft. As I have shown before, various statements made by Adamski and many other contactees seem to suggest the visitors originate from planets where life does not precipitate onto the dense physical planes of matter.[29] As Adamski once stated: "Even upon planets whose atmosphere is so rare that life seems impossible there may be intelligent forms existing – forms having the power of reason such as we possess, but the actual physical construction may be so fine as to be almost invisible to our sight, limited as it is to this particular plane of manifestation."[30]

It is likely that Adamski was familiar with the interpretations put forward by Meade Layne, editor of the *Round Robin*, a "Bulletin of Contact and Information for Students of Psychic Research and Parapsychology", who assumed the space people "materializing" and "dematerializing" for their visits. No doubt his students and visitors at the Palomar cafe will have imposed on him with questions about how Mr Layne's understanding might be squared with his own information. In a letter to Laura Mundo, a supporter from Detroit, Michigan, he explains the difference as follows: "Regarding Materialization and dematerialization. Common sense tells you that if you could do these as the space people are supposed to do, you would not buy a ticket to California or anywhere else. You would just

29 See e.g. Aartsen, 'Can astrophysics see beyond its own limitations?' in *The Sea of Consciousness*, or the page 'Extraterrestrial life' at <www.the-adamski-case.nl>.

30 Adamski (1946), *The Possibility of Life on Other Planets*, p.20.

demat and mat at your destination. The same is true of them. Then what would be the need of building ships? It doesn't make sense, does it. No, they do not mat or demat at anytime, but their speeds make it appear so."[31]

Paradoxically, this is exactly what the Masters of Wisdom, the Elder Brothers of humanity who have gone ahead of us on the path of evolution, are capable of and have been known to do. Several instances of Masters appearing out of thin air were witnessed by students of Madame Blavatsky[32], who was the first to inform the general public about the existence of this super-human kingdom in nature in her work for the Theosophical Society.[33] Contemporary accounts have been published for many years in *Share International* magazine, edited by Benjamin Creme between 1982 and 2016. Likewise, the Space Brothers from the higher planets could equally visit us in this manner if it weren't for the laboratories that their ships are fitted with and that are used for their continuous research of planetary conditions and space itself, as well as to help keep our planet habitable despite our own destructive

---

31 Letter to Laura Mundo, October 27, 1954. In: *The Adamski Documents, Part One* (1980), p.5.

32 See for instance this eyewitness account from April 1881: "...when we were chatting in the above verandah as usual, another Brother, clothed in a white dress, was suddenly seen standing on a branch of a tree. We saw him then descending as though through the air, and standing on the corner edge of a thin wall. Madame [Blavatsky] then rose up from her seat and stood looking at him for about two minutes, and – as if it seemed – talking inaudibly with him. Immediately after, in our presence, the figure of the man disappeared, but was afterwards seen again walking in the air through space, then right through the tree, and again disappearing..." M.B. Nagnath, in: Daniel Caldwell (2000), *The Esoteric World of Madame Blavatsky. Insights into the Life of a Modern Sphynx*, pp.168-69.

33 It is important to note – and cannot be emphasized often enough – that the Society for Psychical Research report from 1885 which denounced Mme Blavatsky as a fraud, was unequivocally retracted in 1986.

habits of polluting, exhausting, and destroying our natural environment. Examples of Space Brothers vanishing into thin air have also been documented, for instance by contactee Truman Bethurum[34] and researcher Cynthia Hind.[35]

In her letter to Riley Crabb, referenced above, Ms Martinelli explains how Adamski's insistence about the physical reality of the space visitors doesn't necessarily contradict the notion that they are not always visible, and says: "I think more people should learn about the real constitution of matter."

To her comment above Ms Martinelli adds that she only goes along with some of Mr Crabb's own notions about materialization and dematerialization, "because I know something about Einstein". Indeed, physics had been making great strides in its understanding of the nature of reality since the 1920s. The originator of quantum theory, Max Planck, wrote in 1944: "There is no matter as such; it exists only by virtue of a force bringing the particles to vibration and holding it together… we must assume behind this force the existence of a conscious and intelligent mind."

The notions of the illusory nature of material reality and an underlying intelligence had long been taught in Eastern philosophy, which gained considerable popularity in the West with the arrival of Indian gurus at the end of the nineteenth century, but their teachings were mostly still veiled in mystical, exotic Sanskrit terminology and concepts. Yet, with the reintroduction of the hidden wisdom teachings of the ages these notions were made accessible for modern Western minds when Mme Blavatsky wrote: "It is on the doctrine of

34 Truman Bethurum (1954), *Aboard a Flying Saucer*, pp.91-94.

35 Cynthia Hind (1982), *UFOs – African Encounters*, pp.138-143.

the illusive nature of matter, and the infinite divisibility of the atom, that the whole science of Occultism is built."[36] To be sure, the 'science of occultism' is nothing to do with black magic, but simply refers to the hidden (= occult) science of the energies behind the processes of evolution.

The wisdom teachings state that above the solid, liquid and gaseous physical levels that fall within our range of vision, there are four further planes of matter, where material particles vibrate at higher frequencies than those on the lowest three planes. In the nineteenth century French physicist Jacques Fresnel had already proposed the existence of 'aether' as the invisible element that fills all space, which Theosophist Charles Leadbeater explained as follows: "That which science postulates as ether is found by occult chemistry to be not a homogeneous body, but simply another state of matter; not itself a new kind of substance, but ordinary matter reduced [i.e. rarefied; Ed.] to a particular state. We may have, for example, hydrogen in an etheric condition instead of as a gas..."[37] When experiments failed to prove Fresnel's theory, science abandoned the notion of 'ether' as a universal substance.

Before quantum mechanics gave us revolutionary insights into the nature of matter Alice A. Bailey, a teacher in the wisdom tradition and a student of one of the Masters of Wisdom, wrote that the atom "can be expressed in terms of force or energy. (...) The word 'substance' itself means that which 'stands under', or which lies back of things. All, therefore, we can predicate in connection with the ether of space is that it is the medium in which energy or force functions... Substance is

36 H.P. Blavatsky (1888), *The Secret Doctrine*, Vol.I, p.520.

37 C.W. Leadbeater (1902), *Man Visible and Invisible*, p.8.

the ether in one of its many grades, and is that which lies back of matter itself."[38]

Based on its calculations of the mass of the Universe astrophysics says it doesn't know what more than 90 per cent of the cosmos consists of and hypothesizes 'dark matter' and 'dark energy' to explain this 'missing' mass. It was the Swiss astronomer Fritz Zwicky who first proposed the concept of 'dark matter', which some scientists now see as a different kind of sub-atomic particle in a 'supersymmetrical' parallel universe "that behaves like an invisible mirror-image of ordinary matter."[39] Dr Zwicky worked at the Palomar Observatory and is said to have visited Adamski three times at the Palomar Gardens cafe, although he publicly ridiculed him.[40]

In February 2022 a groundbreaking experiment at the Karlsruhe Tritium Neutrino (KATRIN) spectrometer in Germany shed further light on the mysterious stage between the visible and the invisible Universe. Reporting on the findings the Max Planck Society writes about neutrinos: "In cosmology they play an important role in the formation of large-scale structures, while in particle physics their tiny but non-zero mass sets them apart, pointing to new physics phenomena beyond our current theories."[41]

---

38 Alice A. Bailey (1922), *The Consciousness of the Atom*, pp.36-37.

39 Steve Connor, 'The galaxy collisions that shed light on unseen parallel Universe'. *The Independent*, 26 March 2015. See: <www.independent.co.uk/news/science/the-galaxy-collisions-that-shed-light-on-unseen-parallel-universe-10137164.html>.

40 Don Lago, 'Messages from Space'. *Michigan Quarterly Review*, Vol.54, No.1, Winter 2015. See: <hdl.handle.net/2027/spo.act2080.0054.108>.

41 The Max Planck Society, 'Neutrinos are lighter than 0.8 electronvolts: Experiment limits neutrino mass with unprecedented precision". *Phys.org*, February 14, 2022. See: <phys.org/news/2022-02-neutrinos-lighter-electronvolts-limits-neutrino.html>.

According to Patrick Decowski, a physicist with the Dutch National Institute for Subatomic Physics (Nikhef), this means there is something special about the particle: "We suspect the neutrino to hold the key to all kinds of things that physics at present doesn't understand." He thinks it could be the source of the elusive 'dark matter'.[42] Although tentatively, science itself now seems to be on the verge of acknowledging that there is no demarcation line, or separation between the physical, visible Universe and that which it can't see.

Interestingly, Adamski may have already hinted at the solution to physics' conundrum when he wrote to Emma Martinelli: "The trouble with the metaphysical setup is that everything in the invisible is labelled 'spiritual' while in the visible it is labelled 'material', but in truth there is neither spiritual nor material – it is all the same." (p.81)

Systems scientist Ervin Laszlo explains that the nineteenth century idea of 'ether' has now re-entered physics as the 'deep dimension' beyond spacetime, known as the 'implicate order', the 'akashic field' or the 'complex plane', which "appears to be a dimension or domain of the physical world beyond spacetime".[43] As Adamski puts it: "All Nature is etheric; whether in a form or formless state… when the word 'ether' is properly understood, you can see it has no reference to spirits or disembodied entities."[44]

All things considered, then, Adamski didn't contradict the

42 George van Hal, 'Wereldrecord: fysici leggen neutrino op "weegschaal" en noteren duizelingwekkend lage massa'. *De Volkskrant*, February 14, 2022. See: <www.volkskrant.nl/nieuws-achtergrond/wereldrecord-fysici-leggen-neutrino-op-weegschaal-en-noteren-duizelingwekkend-lage-massa~b309d9a9/>.

43 Ervin Laszlo (2016), *What is Reality? The New Map of Cosmos and Consciousness*, pp.19-20.

44 Adamski (1957), *Cosmic Science* bulletin Part 1, Question 12.

facts, because the visitors *are* physical, but of a subtler nature than the physical planes in which our own consciousness manifests. Given the intricacy of the issue, Desmond Leslie suggests: "Maybe his mandate was to try and establish only the objective reality of the visitors; a difficult enough task in itself, without confusing the layman with anything remotely esoteric."[45]

Cosmic philosophy – and a new dispensation

Four of the letters to Emma Martinelli that are reproduced here were first published in 1990 as an appendix to *George Adamski – Their Man on Earth* by Lou Zinsstag, Adamski's co-worker in Switzerland until she became "tired of Adamski's articles on cosmic philosophy". She deemed the teachings of the Space Brothers that would help humanity transform the world and make open contact with the visitors possible, "too repetitive and sometimes too abstract".[46]

Like most critics pining for open landings and contact as confirmation of their efforts, Ms Zinsstag missed the point that Adamski made throughout his mission: "The main thing is NOT a question of saucers, my experiences or anybody else's. The main thing is *what is really transpiring in this world. That* is more important than any individual or phenomena that is taking place at this moment."[47] That "main thing" is the fact that the visitors from space are here not to satisfy our personal curiosity, but to assist humanity in taking the next step in the evolution of consciousness, humanity's transition into a new era of good will and universal brotherhood, replacing the current

45 Leslie (1970), op cit, p.259.

46 Zinsstag & Good (1983), op cit, pp.77, 78.

47 Adamski (1955), *Many Mansions*, p.3.

systems that are based on centuries-old habits of competition and greed, which result from a false sense of separation. As Ms Martinelli rightly observed in 1962: "Everyone wants the dramatics, the entertainment of Saucers, but very few want to buckle down and work on themselves spiritually."

No less true today, this observation also explains why the interest in UFOs from the 1960s onward has been limited mainly to looking for – or demanding – physical proof of their existence, while ignoring or discrediting the experiences of the 1950s contactees.

In his talks Adamski repeatedly stated: "[The Bible] tells us that after 2000 years (…) a new dispensation would enter. That the old would pass away and the new would come."[48] But, as he writes to Ms Martinelli in March 1951, "Instead of recognizing the Divine plan working, man resists the change as it is being forced upon him [by cosmic circumstance; Ed.] and he blames this one and that one and someone else for the wrongs which he thinks they are doing – according to his judgment of right, promoted by habits of 2,000 years." (p.68)

In a letter to Gray Barker, editor of *The Saucerian* newsletter, Adamski indicates how the powers-that-be have everything to lose when that crystallization finally comes undone: "Is it because they fear the event which might bring peace and understanding among men on Earth, whereas war has become a financial investment that pays well to certain investors? (...) This knowledge of interplanetary visitors who are friendly to Earth people, and who themselves have learned to live in peace with one another seems to have given new hope to many who had lost hope; and a new purpose to Life to untold

48 Ibid., p.16.

numbers. Is this what those who control the purse strings of the world fear? Acceptance of the reality of Interplanetary Visitors could and would have far-reaching effects upon the present-day economic system on Earth. Everybody would be affected in some ways, but the few would be affected in a far greater degree. I believe they see this and are fighting it with everything they have, at the same time endeavoring to remain conspicuously absent from the scene. (...)

"When people's eyes are turned upward, searching for peace and happiness on Earth, with brotherhood even in a small degree among men here, it becomes more difficult to fill their minds with hatred for their fellowmen wherein war finds a fertile field. And youthful minds filled with challenging thoughts of space travel are not so easily diverted toward the rather questionable honor awaiting in bloody battlefields. Is this what the financiers see and fear? I think it is!"[49]

While many contactees in the 1950s, and several later ones, such as Giorgio Dibitonto[50], concurred with Adamski's expectation of a new dispensation dawning in some form or other, he knew as no other that this is nothing to do with a new religion, but with the manifestation of a new phase in the evolution of consciousness – the inevitable consequence of Earth entering a new cosmic cycle. And although the latter is not yet recognized or acknowledged by traditional scientists, there is now a growing understanding that consciousness, from the most limited in atoms to the most unfathomable in galaxies, is ever evolving towards greater unity, and a number of advanced

49 Letter to Gray Barker, March 11, 1955. In: *The Adamski Documents, Part One* (1980), pp.25-26.

50 Giorgio Dibitonto (1990), *Angels in Starships*.

thinkers, such as professor Laszlo, agree that humanity stands on the threshold of a new awareness.[51] Adamski himself put it thus: "... the world is going through a revolutionary state in which there are many characters. This drama of a revolutionary state is necessary if man is to know the world he has always dreamt about." (p.67) Even a cursory glance today will confirm how accurate was Adamski's assessment of the volatile state of the world, as we see the structures of our old 'civilization' – political, economic, financial – failing.

Hence, Adamski's "repetitive" cosmic philosophy was not only central to his mission, as becomes clear from his letters to Emma Martinelli, as much as from his teachings with the Royal Order of Tibet in the 1930s, but it also shows his understanding was well ahead of his time.

The bigger picture

George Adamski – Their Man on Earth is a facsimile publication of Ms Zinsstag's original, unedited manuscript for the book that she co-authored with Timothy Good in 1983 as *George Adamski – The Untold Story*. This was the first attempt at an unbiased look at Adamski, at a time when he had long been dismissed by most UFO researchers.

Timothy Good introduces the four letters in the appendix to Ms Zinsstag's manuscript, from which he also quoted in his book *Alien Base* (1998), by saying they were "written about the time of his [Adamski's; Ed.] taking of some of his now well-known photographs. Emma admired him and called him professor. He responded by signing his letters to her in

51 See e.g. Laszlo (2017), *The Intelligence of the Cosmos*, or Laszlo (2022), *The Upshift. Wiser Living on Planet Earth.*

the same way, one of the few instances of his using the title at all, and it was entirely unofficial and personal to her".[52]

Although Adamski had been referred to as 'Professor' and used the title himself since his time as a metaphysics teacher in Laguna Beach, it wasn't to misrepresent himself. A biographical note in July 1951, before he became world famous, stated openly: "The title 'professor' was given to him by his students."[53]

The complete batch of Adamski's letters to Emma Martinelli was in the possession of Lucius Farish, who edited the UFO News Clip Service and was the director of the Ozark Mountain UFO Conference in Arkansas. Around 1987 Swedish ufologist and archivist Håkan Blomqvist obtained copies of the complete set from Mr Farish and he was the first to write about them in the AFU (Archives for UFO research) newsletter and later in his blog.[54] Mr Blomqvist has kindly made these letters available for publication in this volume. The first letter, dated July 5, 1950, was retrieved from Japanese researcher Tadashi Takeshima's blog (see page 3, note 2).

As private correspondence Adamski's letters were clearly not a case of public "showmanship", as can be seen from several of his statements and questions in his letters.[55] While Emma Martinelli's letters to Adamski will perhaps never be recovered, they must have revealed considerable understanding on her part as he didn't hesitate to elaborate

52 Lou Zinsstag (1990), *George Adamski – Their Man on Earth*, p.161.

53 *FATE* magazine, op cit, p.64

54 Håkan Blomqvist, 'At Last: A Verdict on Adamski!'. *AFU Newsletter* #31, as reprinted in *Focus*, Vol.3, No.3, March 1988; 'The George Adamski correspondence', blog entry October 29, 2013. See: <ufoarchives.blogspot.se/>.

55 See e.g. note 48 on page 72.

various metaphysical and ethical concepts in his replies, from reincarnation to abortion. Without exception, these show George Adamski's deep insight into the nature of reality and his ability to explain complex esoteric notions and the Laws of Life in the simplest terms.

On several occasions I have suggested that Adamski had been in contact with visitors from space before his reported first contact in the desert, seventy years ago this year. In his letter of November 24, 1951, i.e. written precisely one year before the first report of his meeting with an occupant of a flying saucer in the California desert appeared in the *Phoenix Gazette*, he 'confesses' to Ms Martinelli, whom he clearly addressed as a confidant: "I have never denied communication since I would be denying the action of God". (p.82) Here, I believe, we are given a glimpse of the true depth of Adamski's abilities and contacts, comparable to how some teachers – such as H.P. Blavatsky, Alice A. Bailey, Helena Roerich, and Benjamin Creme, worked with and for the Masters of Wisdom – those enlightened beings who have transcended the strictly human state and who are gradually emerging from their age-long seclusion at this time as the fifth kingdom in nature, to help humanity give expression to our innate oneness in everyday life as we reconnect with the spiritual origin of our existence.

Whereas some detractors claim Adamski was trying to play down the philosophical side of his work at the Royal Order of Tibet in the 1930s, because it "might have potentially harmed his 'flying saucer business'"[56], these letters indicate no such concern or intent on his part at all. On the contrary,

56 Richard Heiden, 'An Adamski Chuckle'. UFO Collective, Google Groups, August 30, 2016. See: <groups.google.com/g/ufo-collective/c/TRu5u-mFgzA>.

they prove that his critics are simply unwilling or unable to look at the bigger picture of his work, which shows that his mission was consistent throughout his life. At the same time, his teaching finds ever more confirmation now that the forefront of science is moving towards a more holistic view of life and reality.

As if to emphasize the need for this shifting perspective, Adamski wrote in November 1951: "Without full vision, man has divided the indivisible." (p.78) Just as this impaired vision lies at the heart of the many crises now confronting humanity – from climate change to mass migration as a result of socioeconomic injustice, it has equally prevented a correct understanding or assessment of George Adamski's mission. And in a world where the divisions are more dangerous than ever these letters underscore its enduring significance.

For that reason his letters deserve a wider audience, not only for the historical context they provide, but also because they constitute a pivotal phase in his mission, and highlight the central thread of his teaching about the Oneness and universality of Life.

Around the time when a growing number of Adamski's staunchest supporters in the US and abroad, including Lou Zinsstag, began to leave him, Emma Martinelli indicated that she did understand and recognize the relevance of his teaching. Writing to Riley Crabb in 1962 she affirmed: "I don't always go along with my old friend, Adamski but more than once, I've finally seen the logic of his theories", emphatically stating: "I am now working solely with Adamski."

Gerard Aartsen
March 2022

Letters to

Emma Martinelli

Acknowledgements

The editor wishes to express his gratitude to Håkan Blomqvist of the Archives for the Unexplained (AFU.se) in Sweden for making these letters available for publication. Also, the aesthetics of this publication would have had much left to desire without the invaluable help and suggestions of Meryl Tihanyi of New York City, USA.

Editor's note:

For this publication George Adamski's letters were retyped with obvious spelling or typing errors corrected, and obviously missing words added in [square] brackets. No other changes to the text have been made. Footnotes add context or comments from the editor.

July 5, 1950

Miss Emma Martinelli
2480 Washington - Apt 203
San Francisco 15, California

Dear Miss Martinelli:

In reply to your letter of July 2, 1950 to Professor Adamski.

The first edition of Pioneers of Space is being sold entirely through Palomar Gardens. It was quite limited in number and is already almost sold out, but we do still have a few copies left. One will be sent you C O D[1], or postpaid if a check or money order accompanies your order. The price is $3.50 plus 11¢ state sales tax, a total of $3.61. It is a beautiful book and we are sure you will receive much pleasure as well as a storehouse of food for thought as you read it.

Concerning Worlds In Collision[2], Professor Adamski is not in full accord with the author either. Take the incident of Venus as a comet lashing at the Earth. That is impossible since Venus is now a planet. The fact is, a planet may become a comet but a comet can never become a planet.

In reference to space ships and saucers, they definitely are real and many of them are of interplanetary origin. Professor has some very beautiful pictures of both

1 Cash on Delivery.

2 *Worlds in Collision* by Immanuel Velikovsky was published earlier in 1950.

space ships and saucers on the Moon. LIFE and TRUE[3] have both asked for some of these pictures and he has sent them but so far neither magazine has bought them. They haven't had time yet to write to us since the pictures were sent out. And have you read the Gold Medal Book - pocket edition - by Donald Keyhoe, The Flying Saucers are Real? This is well worth reading - not metaphysical but a compilation of official records.

Most sincerely,

PALOMAR GARDENS

3 Both general-interest weeklies, *Life* was known for the quality of its photography, while *True* was known as "the Man's magazine".

August 16, 1950

Miss Emma Martinelli
2480 Washington Street
San Francisco 15, California

Dear Miss Martinelli:

Received your most welcome letter and read it with great interest. The analysis you have given is the best I have so far received.

True, I have written a book, but as you should know, I deserve no credit, for all that we can do here in the physical state is to be in tune and make of ourselves willing channels for the light of the Universe to manifest in this lower world of matter. As a personal thought I could never see a writer writing a beautiful story and then ruining it by ending it with hatred, envy, jealousy and even murder. Some way, some of our writers hook onto the beam of all intelligence but before finishing their script they linger back into their personal nature and usually wind up with a personal experience which in most cases is the lowest state of manifestation.[4]

Yes, some of us may be more fortunate than others by being able to stay on the beam once we hook onto it. Some of us attain this state by studying sincerely and truthfully, while to others it is a natural thing to do without any

4 This seems to be in response to a compliment Ms Martinelli makes on the positive outlook and spiritual tenor that characterizes Adamski's account in *Pioneers of Space*.

study. This latter seems to be the case with me. Things just simply come to me and I am able to stay with them until the full revelation has been completed.[5] So I hardly know what to say, since I do not deserve any credit, as I stated before. Yet to the world at large, it may be looked upon in the way you have stated in your letter. There perhaps are many incarnations in the past that have brought me to the present. I realize that. But even there, while I may have earned it, I still do not deserve any credit since all this knowledge belongs to all the universe and not to anyone in particular. The ONLY credit that ANY form may deserve is that it becomes a willing form through which such revelations may come, whereby others may advance.

But speaking of scientists - there are two types: one type is very rare. There are mathematical, limited scientists; and the rare ones, the intuitive scientists and I happen to operate in that field.[6] Your revolutionary discoveries have come mostly through the intuitive scientists. And this is where your young man has fallen down, while he has the spiritual qualities within him, as everybody must have, yet he is in the mathematical bonds, shackled by conventions and traditions; and this is true of 99% of our people. They fail to even see, as you say. Ones that do, see the cause that produces the effect. For an example: the very chair upon which they sit has come out of the ethers, so to speak, into some man's mind as a thought-

5 Einstein seems to refer to a similar process of intuitive understanding when he said: "There comes a time when the mind takes a higher plane of knowledge but can never prove how it got there."

6 Here Adamski foreshadows the distinction between materialist science that regards matter as fundamental, and what is now known as post-materialist science which challenges the underlying assumptions of the scientific method and recognizes the inability of traditional science to account for the emergence of consciousness from matter. See also page 7.

pattern[7]; then it was brought into an effective state where all can see it as a concrete form, as they call it. Yet the chair, as a concrete form, may be destroyed many times while the cause, which is the thought-pattern in the man's mind, can never be destroyed. Men seem to have grown to be psychological giants these days from the effective side of the world, while they remain moral morons from the cause side, which is the spiritual side. As you say, 'this far, and no farther'.

Of course they know not who they are. Therefore they judge all things according to themselves. They are an effect of a cause, so as an effect they are guided by effects about them and firmly believe them to be the real. They seldom take any time wherein they might realize that they are but a thought in motion, governed by a force that is closer to them than their hands and feet. But they see not, nor feel this force because they have allowed themselves to be wholly absorbed by the effects of themselves and the effects about them. They might easily be called sort of zombies: they move but they know not why or what moves them. And no matter how simple an illustration might be, even then they fail to see.

Yet there is not one of these that, if they did not fear to speak the inner feeling, who would not bring forth astounding revelations unto the effective-selves.[8] But, as in all cases, what a man doesn't understand, he fears; so they are in bondage to fear. The effect fearing the cause by not knowing the cause, which is not knowing oneself, for

7 Cf Rupert Sheldrake's hypothesis of formative causation, or 'morphogenetic fields', in *A New Science of Life* (1981).

8 I.e. the person in incarnation in the three-dimensional world.

fear that in case the cause should prove itself the dominant one, the effect would become nothing but a servant.[9] So as a result the ego is built up in defense of the effect, failing to listen or reason for fear of dethronement from a high state of egotistical exaltation to a low state of humble servant.

It takes a brave man to know himself. But while he sacrifices the ego, by doing so he rises as a star in a darkened sky that no longer wonders nor fears but keeps growing brighter by the moment.

Yes, these men desire to know all of this, but at the same time fear what they have[10], which is the effect itself. They cannot be called hypocritical. They are truly desirous of the truth. But the habits of the effective-man prevents them from getting it: yet somewhere, sometime, they must break this bond if they are ever to be free and know themselves for what they are.

Next you have asked how is it possible to be in two places at the same time. That is quite simple, very much like we all experience daily in a mechanical way. A man a thousand miles away talks into an instrument known as a microphone: physically he is at the 'mike', verbally he is in your home or mine, while between his physical self and your home and mine, he is silent - travelling quietly through space on the ether waves. The same is true in the case of television - physically a man will be in the studio

9 A comparison with the esoteric maxim of sacrificing the lower (self, ego, personality) for the manifestation of the higher (Self, the soul) seems in place here.

10 Probably: "fear for what they have".

while photographically he is in your home or mine and maybe a thousand other homes - yet in space he is silent and invisible. Actually, if we were to know ourselves as we should, there is no place to go - here or hereafter. We are already everywhere. The only thing that governs location is the interest that the individual has.[11] It is as the great Master said, wherever a man's heart is, that is where he is. If a man can lose location, there is only one other place he can find himself in - that is in the totality of the universe. You might say then, that everything is within him - while he is the body for everything that is or ever will be. And since there is no beginning nor ending, he is awfully vast.[12]

For an example: if one toe on your foot was hurting, all of your interest would be there and you would not be aware of the rest of your body. Yet your whole body is a thousand times larger than your one toe. But in a time like that, a very small portion of the whole is considered. So it is with man. He may be out in the billionth solar system from here, as far as his physical is concerned, while verbally he is in your home. And he need not be entranced. While entrancement is quite often the case, yet he can function equally the same, if he knows how, as the man at the microphone physically - verbally in your home. And the same is true in vision - physically in the television studio while photographically in your television set.

So your experiences which you have had did not

---

11 In other words, where he focuses his attention.

12 Long before the non-locality of consciousness became a familiar concept through quantum science and consciousness research, George Adamski explains it here in layman's terms.

necessarily mean that the folks or friends were halfway between, although they should know where they are - but they could be on another planet just as easily as this one, and speak just as though they had never left. Yet there is a half-way between, for noone can enter another classroom unless he has earned it or qualified for it after a certain period of time - depending upon what is to be learned. They either go forth to the next classroom, or return back to the earth to learn what they have failed to learn while here. That is where your reincarnation plays its part. If they don't return to this earth, but go to another planet or system, it is transmigration. While learning is incarnation.

For an example: you may know this but to make sure, I will explain - whenever you read my book, or anyone else's, and learn something different than you have known before, that is incarnation.[13]

I hope that I have answered your questions here, so let me say something different now. In this letter I have explained, using illustrations, how one may venture from one place to another, while his physical is in one place and he is in another. That is the way I have written this book. I actually have gone to the places I speak of; I actually have talked to the ones I speak of. To you I can reveal this since your letter reveals much, while to others I keep silent about this. Yet I have used earthly measurements, scientific standards and relative logic and reason for the progression, as an earthly foundation.[14]

---

13 I.e. the new knowledge is absorbed, or 'becomes flesh' (incarnate).

14 In other words, in *Pioneers of Space* Adamski presented his out-of-body visits to other planets in terms of a physical journey by means of space ships.

I am now preparing for another book. Just how soon this will be completed, I cannot say at this time, but there will be another book whose title - so far as I know now - will be GOD AND MYSELF which shall reveal, I believe, much as to who man is and how he functions. Once this is written properly, I believe that much of self-mystery will be eliminated. It will take some time to write it and again some time to publish it, but I will bring such a book out. I feel the need for clearing the mystery is great, especially in the troubled world of today. Should this book become internationally known, it may have the power to reverse the tactics of men toward the brotherhood of man, in due time. Of course, I may be hoping for much, but even a little, I will be satisfied with. Yet the possibility will always be there. How large this book will be, I cannot say at this time for again I will have to go into the total universe to get it. But I am fully aware of it at this time.

I am mighty glad to know that you have yourself comprehended so well the truth of your book[15] and that you are able in your way to promote this truth that more of our earthly inhabitants may realize the beauty of life, if they only once know what is to be done to have it.[16] For even to me as a physical being, this book is like as if someone else wrote it. I can read it time and time again, which I have already done, and marvel at the things that are in it. In other words, I have made the journey to get it. Now in the physical I review it every now and then by reading the book. I believe you understand what I mean.

15 I.e. her copy of *Pioneers of Space*.

16 Adamski is referring here to the realignment with our spiritual nature, and the need to manifest this brotherhood of mankind in everyday, practical terms.

Thank you for your nice letter, and write when you can. I too hope that you do have the opportunity someday to come to Palomar Gardens, or I might have the chance to come to San Francisco before too long, since I have several who are working in that line to get me there for some lectures.

Most sincerely

Prof. Geo Adamski

Professor George Adamski

Images: Peter Bruegemann, Palomar Mountain History Resources.

The Palomar Gardens Cafe along Star Route 76, on the southern slopes of Palomar Mountain, Valley Center, California, where George Adamski, his wife and some of his students settled in 1944: "Here they cleared virgin land and built simple living quarters. Here also they raised a small building to serve as a café for passers-by, owned and operated by Mrs Alice K. Wells, one of Adamski's students. Each member of the group shared in the manual labor that went into this effort, and since [in the aftermath of WWII] heavy restrictions were still in effect regarding materials, anything available had to serve."

Charlotte Blodget, 'Biographical Sketch', *Inside the Space Ships*, p.255

'Highway to the Stars' – Star Route 76, the road on Palomar Mountain that takes scientists and tourists to the California Institute of Technology's Palomar Observatory. The Palomar Gardens Cafe was located off Star Route 76, at about 11 miles from the Observatory.

Image: Peter Bruegemann, Palomar Mountain History Resources.

Visitor brochure for the 200-inch Hale Telescope dome at the CalTech Palomar Observatory on Palomar Mountain, that was completed in 1949.

August 29, 1950

Miss Emma Martinelli
2480 Washington Street, Apt 203
San Francisco 15, California

Dear Miss Martinelli:

Just a note at this time but a more complete answer to your most welcome [letter] will be coming in the near future, maybe the latter part of next week. We are so very busy now at the end of the summer that I am postponing all of my correspondence until after Labor Day weekend holidays are over.[17]

However, there is one thing I wanted to tell you without delaying. Last Sunday a Mr. Donald Kendall of San Francisco came up with the Tanner Tours. He seems to have quite an understanding of many things and was able to follow me very well in the subjects discussed at that time. He asked me many questions and easily understood their explanations. He also bought a copy of PIONEERS OF SPACE. When he gave his address, I asked if he knew you. His reply was 'no'. Then I told him what a fine analysis you had made of the book and how much knowledge and understanding you have. He asked for your address, saying that he would like to meet you and discuss the book with you after he has read it. Feeling that your understanding is of the type that you would have no objection, I gave him your address. When he noticed where you lived, he said he had a feeling that he had either heard of you or had met you, but wasn't sure,

17 When it will be less busy with visitors at the Palomar Gardens Cafe.

however he would make it a point to contact you.

His name and address:

Donald Kendall
1364 Grove Street
San Francisco 17, California

Just one point in finishing. You answered your own question on 'incarnation', 'reincarnation' and 'transmigration'. Learning is endless, as you say, and your analysis as given toward the end of paragraph 1 of page 3 of your letter is absolutely correct. More will come in my next letter.

Sincerely,

Prof. Geo Adamski

Professor George Adamski

September 18, 1950

Miss Emma Martinelli
2480 Washington Street, Apt 203
San Francisco 15, California

Dear Miss Martinelli:

Please excuse this long delay in answering your letter. One never knows what might come up to cause delay in other channels of expression, such as in this case, the answering of your most welcome letter which I had planned on answering several weeks ago.

Maybe you have already met the man, Donald Kendall, of whom you have been informed. He had quite a bit[18], but of course, just like many of us, just stepping up the ladder.

In that particular group with whom he came, a Tanner Tour, there were several persons. One of them was a woman who called herself a 'sister' from the Hindoo institution at La Crescenta, but what she represented does not comply with her behavior. In other words, she claimed to represent a finer quality of spirituality, which is another metaphysical group, and a 'sister' especially should be most humble to represent the humble manner in which the Creator serves unto all men. But far from being such, she was one of these who profess to know and needed no enlightenment. This analysis of her was verified by the driver on his next trip up here and I believe him for he heard the remarks and observed the actions of his passengers during the entire

18 He understood quite a bit.

trip. It also goes to prove even in so small a group, what you said in your letter. They like prestige wherever they can get it, whether from truth or otherwise, so long as they get it for self-perpetuation, which is for selfish purposes. And there are many such as she. Yet I cannot even find fault with such, for they too are groping in the dark only for the light to see, and if patience is exercised with them, in time they too will rise to where the Truth will be their foundation rather than the exaltation of their personal ego.

This particular person left the room in the process of a talk that I usually give to these groups. Going out of doors, she started conversation with one of the ladies of our household, who, during the course of their conversation, told this 'sister' that I could read her thoughts faster than she was able to think them. Her reply was, he must think me awfully rude for getting up and leaving as I did. Yet the information did not help her, according to the information I have received from the driver concerning her speech and actions as the trip continued from here. So, there are many even in beautiful raiment which represents the Divine purposes, according to their accepted symbology, but the Truth they have not. And if they do recognize it, then they do pervert it for selfish purposes. So I can readily understand what you mean when you say some of your material has been too heavy for some of the metaphysical groups. It is not too heavy by any means. Nothing is. But it is sometimes awfully bitter to swallow if one, by that Truth, is proven to be a 'Judas'. For a man may face the most horrible experience of all men, and in the end laugh about it. But when it comes to facing one's self, few, if any, will dare to do so. For no greater criminal has ever lived, or coward, than the one who is made to face himself.

For Truth is something like this.

As an example: A burglar in a house in the dark of the night is moving around freely and relaxed. He may be cautious should he hear footsteps of someone else than his own, so long as there is no light by which he could be identified. But let the light come on and he will struggle to the utmost to hide his identity, even kill to do so, if necessary.

So it is with Truth. It is that type of a light. If it is focused fully upon the hypocrite, the hypocrite will fight with all that is in him to hide what is in him, using all methods to do so.

Yes, this world is wonderful and many in it are gradually stepping forward, while many more are using Truth for self-perpetuation. Even the few who are stepping forward, if the true test comes - which it must to every man - to prove their strength and the truthfulness on which they are supposed to stand, will be able to endure it and stand firm to the end. For some have built their houses on shifting sands, so when storms come along their houses do not stand. While others have built on rocks; such do withstand the storms that come. And so it will be so long as man worships the case he is in, which is his body, doing all for it nor realizing that it has no life of its own but must depend upon the Life of another for its very being. Such ignorance will fight unto the end to maintain itself.

Man for ages, as he is known to other men, has been and is nothing but a shadow, a mask.[19] And this mask takes

19 Here Adamski refers to Latin *persona* (litt.:'to sound through'), the mask

itself for reality. Being an effect, it is guided by effects and knows not the Cause: while it seeks It and hears the words of it and also acknowledges these words maybe, to the fullest extent, yet an effect it will adhere to rather than the Cause. For an effect is like itself: it can feel it and hear it, while the Cause is the opposite – It can neither be seen nor felt. So the effect it worships – and herein lies the statement, the blind lead the blind. For the effect is a blind thing. Could the effect ever be, if the Cause was not master, or creator?

So you can see then, it is not hard for a man to gather large groups, if he deals with effects to satisfy all their effects.[20] While another that deals with the Cause in the largest portion of his act would draw not so many; for he would not respect the effect but would expose it for what it is: which in turn would mean he is stepping on someone's toes. And this, even most of the metaphysical students or groups do not like to have done. Here again we may refer to a statement: Greater is he who saves one soul, than he who takes a city.

And speaking of Duke University, as you say one of the lecturers mentions here and there. Let me tell you of an experience I had with them, if I have not told you already. In the year of 1938, when they came out with the statement about the sixth sense of perception – I called them [out] on that. These were my words, didn't they think that the world was already pretty much confused with the various theories and phrases without adding more to them by adding the sixth

used by actors in ancient Rome to portray a character. In other words, our personality is only a 'mask', a temporary reflection of the real Self.

20 I.e., personal desires.

sense of perception, which was really nonexistent? They naturally got a little huffy and wanted some evidence that there was no such thing as the sixth sense of perception. At that time they were pretty much up against a stone wall, playing with cards only, and did not seem to get much farther than that. Since then they have advanced beyond that wall, yet have acknowledged nothing as to what enabled them to go as far as they have. Neither have they dropped the idea of the sixth sense of perception. Of course I realize it would have been hard for them to retract that after all the publicity had been given it, so I let them go. Again, the blind lead the blind.

This was my proof to them that they were wrong and it is still the same proof today, for Truth cannot change. I told them then that a man did not have five senses, let alone six. Yet some Hindoo teachers teach even more than that. It is no wonder then that so many students have failed to realize what they started out searching for: for they cannot find that which is nonexistent.

Here is the analysis:

Let us say that a man is a by-product of nature - which he is. Therefore he cannot be contrary to nature herself. Nature has four elements out of which all the other elements come forth: earth, air, fire and water.[21] The consistency of each is composed of many others, but these four are governed by Universal law. So, as we said before, man cannot be contrary to these. Consequently, all a man

21 This hearkens back to George Adamski's teaching in *The Invisible Ocean* (1932, pp.12-13), reflecting his training with the Masters of Wisdom in Tibet as a youth.

is, as we know him, a creation of nature and only a four sense human being: consisting of sight, hearing, taste and smell. The sense which is called touch, is not really in that category. If it is to be called a sense, it should be called a cardinal sense.

Example: a man may have X-ray vision, hearing to hear sounds not heard by others, taste and smell the same, but if the sense of 'touch' is not present, he is a dead man as we know dead to be.[22] Yet all of his senses are present and perfect. Only his heart has stopped beating. Otherwise he is a perfect form.

On the other hand, he may lose sight, hearing, taste and smell, but if he has the sense of 'touch' as we know it, he is very much alive and can build empires. Which goes to prove that the sense of 'touch' as we know it, can live independent of the other four; while the other four cannot live independently or in the absence of the sense of 'touch'.

When analyzed as to what the sense of 'touch' really is, it is a feeling or a state of alertness, which finally comes down to the conscious consciousness. And that is the real man behind the mask. So to cultivate this for a finer quality of expression, one must cultivate the feeling. In other words, impose the finer quality of feeling upon the other four senses and thereby the four senses begin to expand into channels in which they have not functioned before. Seemingly, it would be like adding others to the present ones, which is not the case. Instead, the four

22 It may help to understand 'touch' as the three-dimensional aspect of 'awareness', or our subjective experience, without which none of the sensations from the four senses would be registered, such as the 'touch' of vibration on the ear drum, or the 'touch' of flavor on the taste buds.

physical senses have given up their will from the effective side of the world unto the Cause will, which is the Cause side of the Universe.

Illustration: here is a boy who has a violin. There are four strings on it and the boy is learning to play. In the beginning he plays such music that no one can bear it. As he continues learning, he does not add more strings to the violin but he does learn how to manipulate the four he has on it and how to impose upon them the finer quality of his feelings until finally he masters them so well that now he plays such celestial music that you could sit and listen to him by the hour.

So it is with us. Our four senses are like the four strings on the violin. To express the beauty of the Infinite we must impose, with the feeling, upon the four senses of the physical that they too may play the finer melodies of Life by willing themselves to express that which is conveyed by the feeling upon them. Then we really might say, once it is accomplished where all four senses are operating under the guidance of a higher feeling, or finer feeling, which we might say is conscious awareness or conscious consciousness, and what I mean by conscious consciousness, is that the four senses are constantly alerted unto the Universal consciousness and it seemingly looks like we have added other senses to the original four. In Truth we have refined the coarser until it could unite with the higher, the earthly and heavenly and once man accomplishes this, he is a complete man.[23] He is the Oneness he was looking

23 This underscores the esoteric maxim "As above, so below", which is also reflected in the notion that our physical senses are merely the external manifestation of our innate 'extrasensory' abilities.

for. He is the all-inclusive. There are no more barriers for him.[24]

So again we might say that sometimes it is easier to add something that is nonexistent, for people like burdens. The more they have, the more they want. As you said, the diamonds are in their backyards, but they throw them in the trash and look at a distance for that which they had at hand.

Now in regards to the transmigration and reincarnation, you have really answered your own question, as I stated in my last letter. You reincarnate as many times as necessary in any planet and then transmigrate from one planet to another as you graduate from one to a better one. While incarnation means added knowledge as you go forth and there is no end to knowledge. In this, of course, we are speaking in the physical. There is neither reincarnation, transmigration nor incarnation, really, once we become the real.[25]

I hope I have answered some of your questions and I believe that as far as being on the path is concerned, you are very well balanced, and if you adhere to what you now profess, you have help, but keep on going forward for we cannot afford to exclude the world we live in for a world of totality, for that would be division and that is the very reason for man's troubles today - that division. We must not swear by earth that is His footstool, nor by heaven for that is His throne; proving that both must be considered

24 In *Telepathy – The Cosmic or Universal Language* (3 parts; 1958) Adamski elaborates on his analysis of the human senses.

25 I.e. once we have mastered our lower nature and realized the higher Self, or identify with the Source of consciousness.

and worked with.[26]

So again, let me thank you for your nice letter and I appreciate the description of Christ very much. Hoping that when the right time comes I may be privileged to appear in your city for a few of these lectures, or maybe better still, to meet with groups that would gather for a round table discussion. These really bring out more than any lecture. Thanking you again and write whenever you feel like it. Your letters are always welcome.

Most sincerely,

Professor George Adamski

26 In other words, in our quest for spiritual attainment on the path of evolution we cannot ignore or denounce the world of effects since it forms an integral part of the totality of our existence.

In the revised and enlarged edition of *Flying Saucers Have Landed* (1970) Desmond Leslie includes this photo of Adamski's lensless Ihagee-Dresden box camera, "incapable of taking pictures without his telescope, therefore unable to photograph models or nearby objects."
Image: The Leslie Estate.

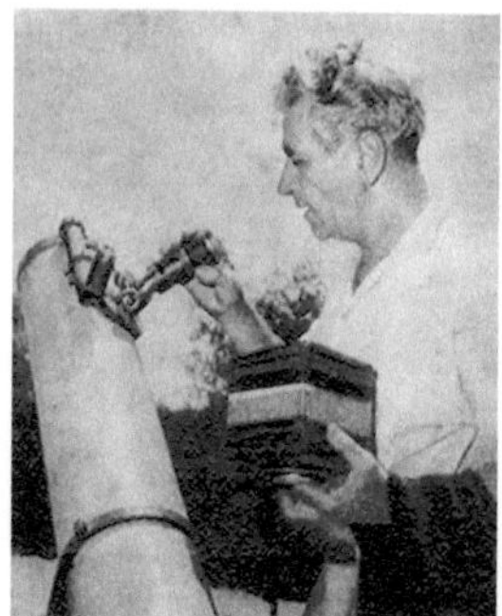

Here we see Adamski mounting the camera onto his 6-inch telescope with which he took the photos reproduced on pages 57, 58 and 74.
Images: *O Cruzeiro* (Brazil, 23 October 1954).

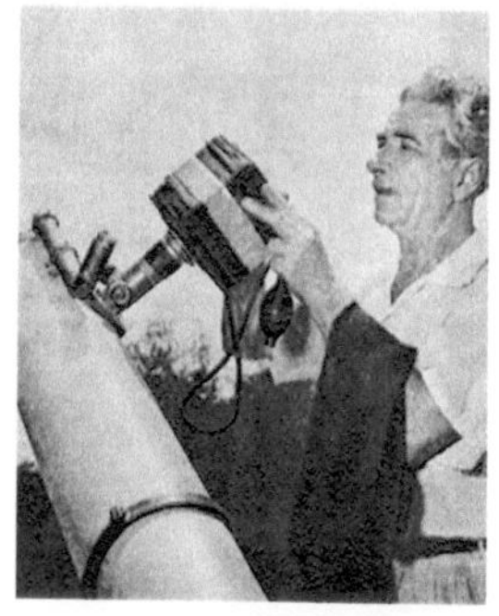

March 13, 1951[27]

Dear Miss Martinelli:

It has been a long time since your letter reached me, also your lovely message at Christmas, and while I should not say this, but in the earthly manner of speaking it is almost shameful to delay in this way. On the other hand, I really have been more than busy.

For one thing, I am conducting a class every other week here at the place. This was rather wished upon me at this time by a number of people who had heard about me, read my book and yet wanted more light and understanding about many things. They needed the help that I could give them so I undertook this class program even though I am very busy with a number of other things, but there is no time like the present.

Just now I am about half way through in the writing of my second book. Of course this book is going to deal with something entirely different than the first one. Its purpose is to alert earthly man to the manifesting of the universe as it is now being revealed to him through the presence of saucers and space ships in our own atmosphere. It does carry a portion of fear, due to the strange things that have been taking place in the last few years, like lost aircraft and other strange manifestations in space. Then there is the constant vigilance of saucers and space ships over military proving grounds for which there must be a reason – and this

27 This letter was incorrectly dated 1950. See also the reference to 1951 as the correct date on page 68: "For remember, we are (...) in 1951 according to man's messing up of the time records."

reason I am working with. It will be all fiction but based on fact, and might open up the minds of earthly men; whereas nothing else probably would ever be able to do so. And by the looks of things at present, I may have a better chance of having this book published for me by a company than I had with the first book.

Just three weeks ago I was a guest in Frank Scully's home where there also was a group of foremost scientists. Here many facts were discussed that, to the average man, would be unbelievable. Also at that time I met the man who was responsible for getting Frank Scully's book "Behind the Flying Saucers" published for him. This man asked for my script as soon as I have it finished. Included in this second book will be eight to twelve pictures of flying saucers and space ships taken through my telescope. One in particular the military has identified for me as being a big space ship only about 5,000 feet off the moon and casting its shadow on the moon. Since the moon is 240,000 miles from the earth, this ship is inter-planetary without a doubt. And I am still trying for better pictures.

By the way, this picture and four others[28], with a story explaining them, which I have written, will be published in the FATE in the very near future, I haven't yet received word just which issue.[29]

Now back to the meeting with the scientists, three weeks ago. Some of them had just returned from Las Vegas and the nearby proving grounds where they had been studying the results of the five bombs exploded there not

28 Reproduced here on pages 57-58.

29 Titled 'I Photographed Space Ships!', it was published in the July 1951 issue.

Caption in *FATE* magazine, where these photos were originally reproduced upside down: **"Sequence of moon 3:30 to 4 a.m. May 6, 1950, shows strange object suddenly appearing over its face. Note small shadow cast by object appearing in photograph on the right."**

[Note also how the ship can be seen approaching as a dot of light in the lower left corner of the second photo; Ed.]

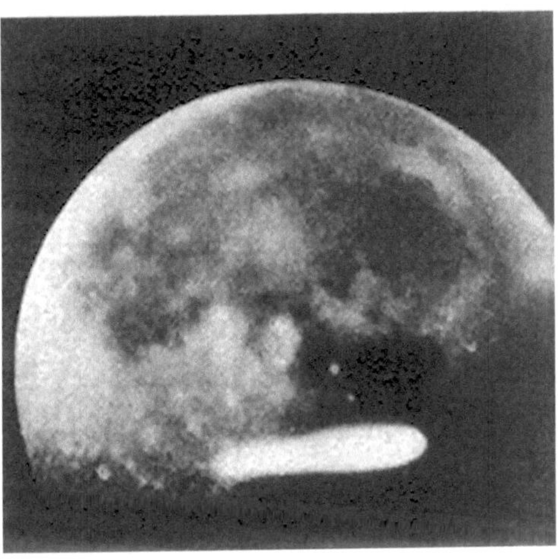

In this enlargement of the third photo the shadow appears as a black area, darker than the dark patches of the moon's surface, right of center above the ship.
Image: *O Cruzeiro*, 23 October 1954.

From the article:

"Reproduced [on page 69 in *FATE*] is a series of three pictures. Notice in the first and second that there is only the picture of the moon, while in the third there is also a picture of a monstrous ship. Watching through my telescope during the early morning hours of May 6, 1950, I observed this bright object moving at terrific speed through space and it seemed to me it was heading straight for the moon. (...)

"Early in November, 1950, having heard much of me and my pictures, a group of high ranking military men whose names and branch of service I cannot give since they are still in active service, came to my place to see my pictures and to question me concerning them; also to look through my telescopes. All of these men have been well trained in aerial photography and in reading aerial photographs. When I handed them an enlargement of this picture, one of them remarked, 'Why, that object is casting its shadow upon the moon.'"

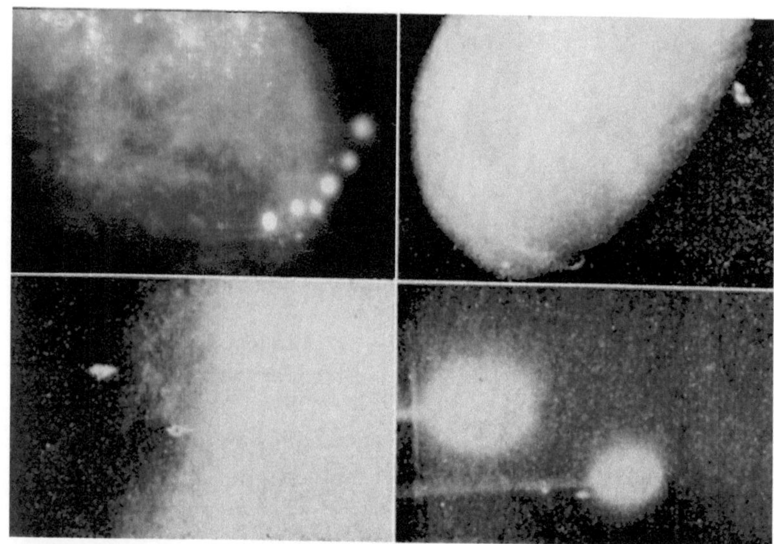

Caption in *FATE* magazine: **"Upper left shows chain of objects between 10,000 and 35,000 miles from the earth. Upper right, estimated distance is 50,000 to 210,000 miles from earth. Lower left, objects are near moon. Lower right, objects in space. All observations on May 27, 1950."**

"The other pictures were taken during the early portion of the nights of May 27 and May 29, 1950. The reflections of the bodies in these pictures are round instead of long. This naturally gives them the appearance of being those objects which we have named 'saucers'. Note the contrast of their reflections to that of the moon.

"In one of these pictures [top left] can be seen seven saucers in what appears to be a formation. (...)

"Another picture [top right] shows four saucers in almost square formation, not too far from the crater edge of the moon's photograph. In this case the moon was much more exposed than in the previous pictures. (...)

"The third of these pictures [bottom left] was half exposed to light and I caught only a portion of the moon in this one, but there are definitely two saucers in this photograph – one at the edge of the moon while another is following the first toward the moon. (...)

"The last of these pictures [bottom right], showing two very luminous balls of light with trails and two smaller objects which appear to be farther out in space, was not taken with the moon in the background because they were not in line with the moon when they flashed through space and I succeeded in getting a shot of them."

Source: *FATE* magazine, Vol.4, No.5, July 1951, pp.64-74.

too long ago. Their findings and the things which they have witnessed are such that one does not even care to talk about. But I believe man on earth is playing too much with the universal laboratory of chemistry while he doesn't know enough about it to be given a free hand at it. Yet that is the situation. Only something unpleasant in the form will awaken us all to the realization that we had better learn more before playing around too freely in Nature's laboratory of chemistry.[30]

That same night a few scientists left here in a special plane for Scotland where a large space ship was said to have landed with forty men, all six feet tall or better, and all alive. These men are now guests of that country. Yet our papers have said nothing about it. When these scientists return from this trip, a group of them are to meet at my place and I will learn more about this strange ship, if it is true.

One of the men has some metal, glass and a radio from some space ships that have landed in this country and I have been told that these parts will be brought to me here so that I can examine them myself. I have been promised this will be within the next month or so. After that I can tell you more, but if the people could only know what is already taking

30 This comment foreshadows the Space Brothers' concerns about the nuclear fission technology that is applied in nuclear tests, weapons and power stations, as disclosed in the first newspaper report about Adamski's contact in the California desert on 20 November 1952 (see page 99). In *Pioneers of Space* (pp.84-85) Adamski already quotes a scientist on the Moon who says about nuclear power: "...what you Earth men have now is dangerous to work with, for you haven't got anything except wild, uncontrolled power." Robert Hastings' documentary *UFOs and Nukes: The Secret Link Revealed* (2010) about nuclear weapons shutdowns coinciding with UFO sightings also seems evidence of their concern.

place in this world of ours in the field of science as well as inter-planetary progress, they would right now believe themselves living in another world.

Now to answer some of your questions, and let me say your letter of September 26 was excellent and I must say you have a broad understanding, with the exception of a few minor yet important corrections.

Take for instance your statement about Saturn, Capricorn, Stalin and Jesus. Remember that no man has made the earth upon which man lives, nor has man made himself as he is, a physical being. And this is true of other planets as it is of ours. Therefore the only one who could have made all of these planets, including all forms upon them, was the Divine Force of the universe. And Himself, being the holiest of the holies, there is none holier. Therefore all of His creations, regardless what planet it be, or what form, is as holy as His holiness. And when we base a judgment against one another or against any given planet, we are basing judgment upon His holiness. There could be no greater temple of holiness than the temple within which man lives.

The Great Book says "swear not by heaven for that is the Lord's throne, nor by earth for that is His footstool." The Saturn myth is based on superstition wherein truth is not present. And to have the Truth, superstition cannot be tolerated. Besides, the planet Saturn if anything, as stated in my book, is the planet of justice, the planet of balance in our own system. That is why the rings. She has nothing to do with the picturesque character in red tights.[31]

31 I.e. the devil.

As far as men of this world are concerned, ours or others, we are all children of the same Father, regardless of color or race or nationality and again I take this stand – "let the one without sin cast the first stone." I know of no innocence. In other words – if I am to be one with the Father, I cannot do any different than the Father – that is if I am to have the Father's truth, for He is not a respecter of persons. Men misbehave in all phases of life, but they and they alone must balance the scales for nature accepts nothing unbalanced.[32] The law is perfect. Sometimes, through much suffering, a thought of hate, a thought of murder is equal to an act in the bosom of nature. So unless we see these things and govern ourselves by the absolute Truth, we ignorantly and unwisely continue judging others – even as do the masses.[33]

I hope this does not sound too rough but to the excellent state in which you already are, this can be a tremendous stumbling block unless you free yourself from it. I do not support astrology – not because I am an astronomer, but because from the beginning man has been given dominion over earth, air, fire, and water and all things upon the earth – if he would only learn the Truth. Tho it speaks of this particular earth of ours, all planets are earths like this one; therefore no planet has any governing power over consciousness, which is man.[34] It is consciousness that governs all – unless man, through

32 Cf the Law of Karma, which is not about retribution, but about restoring harmony wherever our actions or thoughts disturbed it.

33 Adamski emphasizes here the need, as taught in the Wisdom teachings, to control our thoughts and emotions, in order to create right relations and to progress on the path of evolution.

34 In other words, man is the ruler of his own fate, rather than the plaything of forces beyond his control.

ignorance and superstition permits himself to be enslaved by ignorance and superstition. And since man is a mystery to himself, not knowing himself as he should, he therefore has no confidence in his own being. Consequently he uses these various props to support himself, instead of the knowledge he should have, and they work for him because he makes them work by placing his faith in them.

Now in reference to Mount Shasta. Mount Shasta itself is a remnant of Lemuria and it does carry the vibration of that civilization – a civilization that had risen very high at one time, with a good many master minds that could do things that men of today cannot. So I can readily see how you could have been helped, yet hearing nothing and seeing nothing. And who knows – if one was to go into your past lives but what you might have been a part of that civilization? Yet the many things said about Mount Shasta, which again are bordering on the lines of mysticism, have been promoted by metaphysical slickers.

As far as entities are concerned, I can't understand how people fall for such – like in some spiritual planes where they are called 'guides'. If my father knew nothing of medicine while in this world, and especially nothing at all about the structure of my body or his own, how then could he advise me about the care of my body or its needs, when he has not been gone out of this world maybe more than six months? Because he left this physical body doesn't signify he is a master. He still must learn by the process of growth like he would have done had he learned here – maybe a little faster, but again that too depends upon how fast he wants to learn, for he is still a free personal[35] consciousness.

35 I.e. individual.

Although that doesn't mean that the higher forces cannot be contacted; for there are forces that can help us grow faster here, to be ready for there.[36] But I am, like you, speaking of your friend. One shall not allow himself to be a channel for most anything and everything only to be the center of attraction especially, for that is egotism of the highest degree and such can never contact the higher. It is quite unfortunate that in this channel of universal communication so many allow themselves to go to such low depths just to be somebody. I am not judging here, but I am expressing the facts as they are; for it is true that we must help those beneath us, but we must be higher before we can do so. Man cannot pull a man out of a well, by himself being in the well. Rather, he must be above that he might lift up the one he would help.

You see, my dear friend, I am 100% for the truth and therefore the many chaffs must be cast away before the heart of the kernel can be found. One thing to remember, there could never be a counterfeit without the real being first.[37]

Yes, one may travel at will [to] any place in the universe without taking his physical body since the physical is not man, but rather the house of man. Man, himself, is what is termed to be the spirit. So let us illustrate in short what the spirit actually is. The spirit is nothing more nor less than the conscious consciousness which is called man and is no different than the conscious

36 This shows Adamski made the distinction between astral psychism and mediumship, or "communicating with spirits" on the one hand, and contact with higher beings or forms of consciousness, e.g. through telepathy. See also his definitions and explanation in *Flying Saucers Farewell*, pp.111-12.

37 By analogy, would there be flying saucer hoaxes without actual flying saucers having been reported first?

consciousness known as the Divine Father of the universe – and it is everywhere. The only difference between It and man is that that the Father's conscious consciousness is universal while man's consciousness is his interest[38], or focalized conscious consciousness, made so through his physical being.[39]

As Jesus said, "Where a man's heart is, there he is also" which means the consciousness. And let us quote some more: "Where two or three are gathered together in my name, there am I also", which means consciously conscious of the universe and all things within that consciousness naturally would be present – there is nothing outside of consciousness.[40]

To further illustrate this, and I will write a book along this line[41] after I finish this one I am now working on: you see, neither you nor anyone else, not even a blade of grass or insect nor any form known to man today, even the mineral, can live as it does in the absence of the air it breathes. It is this air that is actually the consciousness and it is everywhere; it matters not where the form may be. And so long as it is passing through that form, that form is intelligently acting or expressing. Once the air ceases to pass through it, the form is no longer known as alive but instead it is called dead because it no longer speaks nor acts. In other words, it is not conscious. We might say then that the real man, known as a soul consciousness or

38 I.e. his attention, or identification.

39 In other words, self-consciousness focused in the physical form.

40 Not only is this the view increasingly held by 21st century science, but Adamski already taught this in his first publication, *The Invisible Ocean* (1932).

41 This may refer to either his course *Telepathy – The Cosmic or Universal Language* (1958) or his book *Cosmic Philosophy* (1961).

whatever is its name, is passing through the body making the body possible by doing so - never for a moment stopping in it. For should it stop to linger a while within that body, that body would die, as we know death; for no man can hold his breath for a period of time and live. He must constantly inhale and exhale to live as a physical being. So this then, called the breath of life - the air we breathe - is actually passing through the house of clay by the moment. Once it quits passing through it, then the house of clay crumbles. And since this air is everywhere in the vast universe - some cases in different degrees than we find it here, but it is still the air by which all forms are supported - including planets - then where is there a spot in the universe that man is not already present, or can be consciously as he chooses?

Yes, the consciousness had to be before the building, or house within I live, could have been built. And that is also true of the physical body. I can build many homes in various parts of the world. The fact that I am living in one does not limit my building others in other places and visiting each as I desire to do so. Nor is it necessary for me to take any one of the buildings with me in order to go to the others. Just so, it is not necessary to take the physical body along as one travels through the universe.[42] To thus travel, it is only necessary for one to be conscious of one's self, which is the individual - not the body - for as a conscious being I will always have a pattern in my mind for the next house that I wish to build.

Notice, each time we build a home to live in - not

42 See page 7 for what e.g. physicist Russell Targ says about projecting one's consciousness for the purpose of remote viewing.

meaning the physical body, but a house - we try to improve upon it, put something new in it that was not in the old; a different style to conform with the territory in which we are building, and so on. That is true wherever we go, whether in the material, physical or universal.

Once we leave this physical body we will desire an improvement in the next physical body that we take on whether it be on Venus, Mars, any other planet in our system or even in some system other than our present one. The physical body we are now living in would not do us much good elsewhere - we would only have to repair it. So it is much easier to build a new one. And the conscious consciousness is the builder of all forms without limitations and the breath of life passing through them all, everywhere without end.

And do not forget, as I stated before - judge not, lest ye be judged. Let the one without a sin cast the first stone. Remember always that the very breath of life, or conscious consciousness, flowing through your body and mine, also is flowing through theirs whom we are inclined to judge. For that conscious consciousness, known as breath, actually is coming from the mouth of the Divine Father into the nostrils of all forms; even as it was given unto the first man who was made from the dust of the earth in the likeness and image of the Divine conscious consciousness.[43] Then the breath of life, as conscious consciousness, was breathed into the nostrils and it became a living soul - known as man. So you can see, when we judge another, we are judging the Divine consciousness that is still flowing out of the

43 For an esoteric interpretation of this metaphor, see e.g. H.P. Blavatsky, *The Secret Doctrine*, Vol.II, *Anthropogenesis*.

mouth of the Divine Father unto all forms and without it, no form can live.

So I hope this explains all. I thought it was quite necessary to have these few points cleared since you have so much already and a little mistake can hinder one for a long time.

Now in reference to the moon on the night of the eclipse. Yes, I watched it through my telescope. I did not see anything unusual, although imagination can play a lot of tricks on a person - the earthly type of imagination, I mean, not the one that comes from a Divine source. For the first likes to picture things like unto itself and in which it is sincere in doing so. But once we know ourselves, the physical is made to look upon the real picture and not promote its own.[44]

In the finals let me say, the world is going through a revolutionary state in which there are many characters. This drama of a revolutionary state is necessary if man is to know the world he has always dreamt about. But this physical habit-like effect is bound so by habits and fears of the future that it resists the change, even if it takes its own life to do so - which could have the result of man's annihilation from this earth planet. I would rather say here, from the way I can see it - it is a Divine plan and these characters that we sometimes consider so bad are nothing but actors to bring forth the result of this Divine plan. And unless man has faith in his Creator, he will have to endure suffering such as, if he had faith, he would

44 That is, the personality is brought under the control of the soul, which we then know to be our true Self.

never have known. For remember, we are in the 2012 year, according to the universal calendar, and in 1951 according to man's messing up of time records. In other words, we are already 12 years in the new dispensation. But before we can have the total consciousness of it, the old will have to be removed. And it is not easy to remove the old if the old has become a habit.

In most cases, as we see it now, suffering or pain will be the medium through which this will come.[45] Yet it is not necessary if man had faith in the Divine plan and let the Divine handle it without man's resisting it. For this has been promised – that at the end of 2,000 years from the birth of Christ, a new dispensation would be entered. And the 2,000 years were up in the fall of 1939 for what is known as the Star of Bethlehem appeared at that time over the Pacific coast. And it takes 2,000 years as a cycle for her to come into the vision of man – every 2,000 years.[46] But man has lost sight of the Divine time when he messed up his records and changed his methods of figuring and keeping time. He not only lost sight of the Divine timetable; he also lost his faith in the Divine Father. Instead of recognizing the Divine plan working, man resists the change as it is being forced upon him and he blames this one and that one and someone else for the wrongs which he thinks they are doing – according to his judgment of right, promoted by habits of 2,000 years.

So it would behoove a man to be a little cautious

45 The mounting crises of inequality, pollution, environmental destruction and climate change that humanity has been ignoring to its peril, are now [2022] coming together in a perfect storm, underscoring the accuracy of Adamski's vision and understanding.

46 See also his elaboration on page 83.

of judgment - in so far as judging things which are not altogether of man's own making. The only thing of man's is his resistance against the change which is causing much worry and suffering in this world today; and no man nor nation can be held fully guilty, but all are equal in guilt, if there be any guilt, for separating themselves from the guidance of the Divine intelligence. For many things that God gives to us become curses when He is ready to give us something better to replace the old - not through any fault of His but through the fault of man's habitual possessiveness. You cannot have the pie and eat it too, as the old saying goes.[47]

I hope this clears quite a few points mentioned in your letter, even though it is a little late coming. May you have the best of progress in all your ways and the guidance of the Divine Hand for your future.

Yours sincerely,

Prof Geo Adamski

Professor George Adamski

47 Here Adamski anticipates the efforts of the world's reactionary forces to thwart the manifestation of humanity's Oneness as advocated by the Space Brothers according to many contactees, e.g. by securing world peace through sharing the world's resources towards socioeconomic justice and freedom for all; by regulating corporations to protect basic human rights and the planet; and by allowing the United Nations to take its rightful place as the parliament of nations. See e.g. Aartsen (2015), *Priorities for a Planet in Transition – The Space Brothers' Case for Justice and Freedom,* chapter 2, 'Earth's cosmic isolation – A self-imposed confinement'; or Aartsen (2022), *UFOs and the Pioneers of Oneness,* chapter 3, 'The algorithms of evolution'.

September 30, 1951

Dear Miss Martinelli,

Just a line or two to let you know I haven't forgotten you, but had to get the script of the book out – the one I told you in my last letter that I was working on. It is in the hands of an eastern publishing company now for consideration. I am expecting to hear from them any day, one way or another, although they did ask for it when they learned I was working on it. My article and pictures in the July issue of FATE flooded me with letters of inquiry from all parts of the nation, besides people have driven up from long distances to talk with me. Also I have given a few lectures and of course the business requires a certain amount of my time.

We have rather expected at least a short note telling that you were on the way to visit us here at Palomar Gardens, although maybe since you didn't hear from us you took for granted that we might be too busy, or something. Of course, like most everybody, we are always busy, but your visit would have been most welcome. We should have answered you immediately and let you know that we did have available accommodations. So if your plans were upset because of our negligence, please forgive us.

Answering your question concerning the big space ship with its passengers that landed in Scotland, I have recently been informed that the ship did not land in Scotland, but instead in Australia near Sydney. The number of men was correct – 40 – and they were all around 6 feet tall. According to my informant these visitors just left for their home planet about 2 months ago. They were flown in 2 earth ships from Australia to Scotland where the conference was held, although nothing has been revealed to me as to what was discussed there, nor from what planet they came.[48] The person who gave me this information I know very well. He made a special trip up to see me and tell me. He also is a friend of the pilot of one of the planes which transported the visitors from Australia to Scotland and back to their own ship for their return home. There was no base in Scotland large enough to permit the landing of any ship as large as theirs.

Under very interesting circumstances I had previously been told of a big space laboratory 1400 miles from Sydney, in the same state in Australia, that has now been in operation for the past three years. I was made to understand that space ships could be landing there as well as a communication system could be going on through this laboratory between earthmen and spacemen with even a possibility of inter-planetary communication.[49] It wasn't given to me as a definite fact, but as a possibility from which I was to draw my own conclusions - and I was offered a job there if I was interested. You can make your own deductions as well as I.

As far as space ships belonging to our planet is

48 Detractors and Adamski bashers should ask themselves why, if his accounts were made up, he would not have spun a wonderful yarn here to make himself look the more interesting to an obvious fan like Ms Martinelli.

49 Timothy Good rightly noted that Adamski's information about a possible extraterrestrial base in Australia "pre-dates any publication relating to the existence of alien bases on our planet" (*Alien Base*, p.253), which have since been reported around the world.

concerned, there is no doubt in my mind that we are working on something of the kind and even have some flying around on test flights, but I doubt very much if they are able to go out as far as the Moon as yet.[50] And just here a certain coincident might well be considered, relative to space ships and the recent visit from men of other worlds. Under date of September 3 the Los Angeles Times carried an article revealing that a group of nations, including the U.S.A., were in confab in London along the idea of building up space ships - which indicates that something must have been revealed to urge these nations to unite in knowledge and undertaking of such a program. In other words, it seems as though we have allies in another world - but not from a nationality viewpoint, rather as an Earthly project.

Earlier this year I was successful in catching some very fine pictures of space ships and saucers, with clearly defined outlines and indications of movement plainly visible. These are both daytime pictures and night. Some of them were also caught by the telescopes on top[51] but such information from them cannot be expected, at least at this time. They are working in conjunction with the military and are not free to release any such facts. However, because they did catch some of the same pictures which I got, I had mine copyrighted. As yet none of these have been published, but in the near future I hope to have some extra prints made and available to the interested public since I have been receiving innumerable requests for such pictures.

50 The late Apollo 14 astronaut Edgar Mitchell asserted in a radio interview: "I suspect that in the last 60 years or so there has been some back-engineering (...) but it is not nearly as sophisticated as what the apparent visitors have." (Interview with Nick Margerrison on Kerrang! Radio, 23 July 2008.)

51 I.e. at the CalTech Palomar Observatory on Palomar Mountain, see page 42.

The clear outline of a saucer-shaped craft is visible against the moon, in this photograph taken by George Adamski on June 6, 1950. When inverted (below), the image bears a striking resemblance to the unidentified craft in the videos released by the Pentagon in April 2020 (inset). Even at the much lower resolution and without expertise or digital enhancement, a similar force field that surrounds the craft in the Pentagon photo seems discernable around the craft in the inverted photo.

Original image: George Adamski, *O Cruzeiro*, 23 October 1954.

Image: Department of Defense/ US Navy.

All of the first edition of Pioneers of Space is gone and I am getting numerous orders for the book, which I can't supply. When the second edition will be available, I don't know, but I believe it will be published by a firm this time instead of me having it published myself.[52]

Just as soon as I get a little more time I am going over your letter carefully and answer the questions you asked or comment on the information you have written and there is much of interest that I believe you would enjoy, but right at this moment I have much work piled up before me. Please forgive my delay in answering your most interesting letter and write whenever you feel like it. Your letters are most welcome even though I am slow in answering.

Most sincerely,

Prof Geo Adamski

Professor George Adamski
Star Route,
Valley Center, California

52 Any concrete plans for a second edition would have been superseded by the events that unfolded from November 1952 onward, leading to the publication of *Flying Saucers Have Landed* (1953) and *Inside the Space Ships* (1955).

George Adamski seen near (left) and inside (below) the small observatory dome that he and his group constructed at Palomar Gardens to house his 15-inch telescope.

Image: history.denverlibrary.org

The dome was "designed in a way which enabled him to study the skies for hours on end, protected from inclement weather. The smaller 6-inch telescope was mounted out in the open".
Charlotte Blodget, 'Biographical Sketch', *Inside the Space Ships*, p.255

Image: *FATE* magazine, Vol.3, No.6, September 1950.

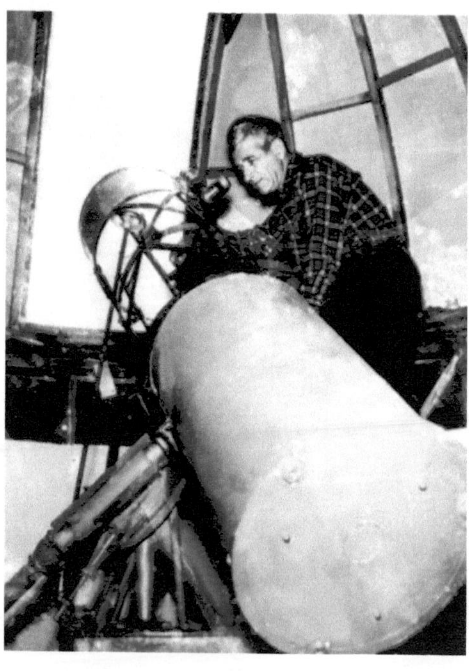

November 24, 1951

Dear Miss Martinelli,

First I will take up the points in your letter of March 25 and endeavor to answer those questions.

Speaking of your little plant - this philodendron whose leaves you had given a milk bath - true, they understand us, while we dumb clucks don't understand them for there is a universal language which is conveyed through a channel of feeling. And 'feeling' is conscious consciousness. All form life is conscious in one way or another, for that is its life.

Now in reference to ships disappearing - the script of my second book is now in the hands of the Henry Holt Co. I haven't yet heard whether they are going to publish it or not. Frank Scully, when I last talked with him, thought they will. Yes, we are in a transitional period and some of it might not be too pleasing to some, but to those who understand, it will be pleasing. The veil between us and the universe is gradually being pushed apart and those of us who have patience as well as vision will witness things which were undreamt of at one time.

Speaking of God and man - spiritual and material - there is neither of these in the absolute truth. They are just manifestations, and at times they are but momentary ones. For an example, could the Creator create anything lesser than He Himself is? I say 'no'. So the very earth on which we live is as holy as any holy could be, regardless of where you might go in the universe: and man - the physical

- is equally as holy. When he realizes this, then man - physical matter - realizes for the first time that it is being used only as the clay in a potter's hands, subject to change as thoughts change; with each change contributing something very necessary to the total universe. Without full vision, man has divided the indivisible. He calls one part 'high' - the other part 'low', thereby seeing a gap between the two and fails to see the continuance of the whole. For instance, one must have feet to walk, yet it is his head that governs the feet. One is high and the other is low, but each is essential to the other and both are one: governed by one and the same power.

It might be said that the 'high', or that which some are inclined to call 'spiritual', is all universal; while the thing called 'material' is only [a] momentary manifestation in a particular section of the universe. Such manifestations are everywhere, making up the total actions of the universe; or permitting the totality of the universe to express simultaneously, as it were, throughout His whole body as a feeling - pleasantly or otherwise.

Unless this is understood, divisions in the minds of men can easily be formulated - which they have been. So it might be said here, that matter has been lost to its own identity by feeling itself separated from the highest, yet it is necessary in the universal picture for it to be where it is at all times. The foot of a man may say, "why in the world did he put me in this mud and get me all dirty," yet it is very necessary for the foot to go into the mud to carry the rest of the body (the 'high') across the pond to dry land where it was going. Both ends of the body partook of the same experience at the same time, for while the foot was in the mud, the high (head) realized it, or was well aware of it, and in turn the whole enjoyed the dry land. Right here it might be pointed out that a mutual interblending continuously took place for the manifestation to be complete.

And so it is with life as a whole. There is continuous interblending between the visible and the invisible - the high vibration and the low - but never a break: no gap: no division. So you see, everything is necessary and there is no fault to be found, even though some experiences may not come up to some of your desires and standards at times - but our standards are self-opinioned rather than universal. All are set up by judgments and indicate division.

No! Man is as holy as he will ever be, for the Creator could not create anything lesser than Himself. And His whole creation is equally as holy as He. What we must do is to realize this and then live it, instead of living in division. Then we will be truly living in heaven, serving and expressing according to the universal plan.

You see, when we speak of matter we are speaking of the spiritual in a lower state of manifestation which is very necessary. A diamond could never have been a diamond with all of its brilliance without having first gone through its slimy lower stage. But all the time it was going through that stage it had within itself the potentials of its purity and beauty. And matter - the slower rate of vibration - must be endowed with a certain form of intelligence to obey higher intelligence, yet there is neither higher nor lower, for a child is a man and a man is a child. Yet what we call a full grown man will talk down to a child man, making it understand in a childish state. So the same is true with matter - it is like a child of the spirit with the spirit likewise being the child.[53] As previously stated, the head doesn't go into the mud, but the foot does, yet both are one and the same - one cannot be without the other and fulfill its mission, for while they appear to be two, they are just one.

Again it might be said that all there is is the Creator and no such thing as man. This Creator expresses Himself in the various stages from one end to the other, regardless what the opinion of matter might be. He is matter Himself. This may be a little hard at times to conceive, and words are limited to convey the exact truth or the exact picture, but that is because our minds are in their infancy and the mind is a sort of link which connects the All-Intelligence with the various manifestations - or the manifestations with the All-Intelligence - for manifestations are always of that slowed-up rate of vibration called 'matter'.

When I say minds of infancy I mean just that because the average mind today is less than 6 months old, although it may be within a physical body 60 years of age. Let me explain how I obtain such a figure: the average person sleeps 8 hours, works 8 hours in the business field where he has no opportunity to think of or for himself, leaving 8 hours for himself. Of these 8, he will use 4 definitely going and coming to work, dressing and meals. He has 4 hours left in which he may pick up a paper or book and read someone else's thoughts, or seek entertainment of one kind or another. Should he sit down for 15 minutes wherein he does

53 In other words, just as matter would be hapless without spirit directing it, spirit would not reach its full potential without matter to manifest itself.

his own original thinking without influence from anyone else, he is doing very well; but it is this kind of thinking that actually develops his own individuality and his own mind. Now if you take that 15 minutes of each 24 hours over the course of 60 years of physical life, his mind is but a babe's while his body is of age. So, considering all these facts - what man really understands - I believe he is doing a pretty fair job with his little development. Of course, the more matter is being used, the more sensitive it becomes, or endowed with consciousness of perception: or by being a servant, the finer matter becomes, just like the diamond. Yet in the final analysis, matter and the universe are but one and the same thing - the only difference lies in the extreme manifestation.

The blade of an electric fan may be a visible obstruction for an object behind it, yet when the blade goes into a high state of action it becomes invisible to sight, but that doesn't mean the blade isn't there. The only difference now existing is that previously it was in the still state wile now it is in a state of high activity, yet it is the same thing. The trouble with the metaphysical setup is that everything in the invisible is labelled 'spiritual' while in the visible it is labelled 'material', but in truth there is neither spiritual nor material - it is all the same.

I do not find any fault in relationship between planets like conscious relationship. It is now known that birds have migrated from other planets to this one, and from this one to others. This is not known to have taken place in this immediate territory, but in this world.[54] And no child

54 I have not been able to determine the source that Adamski seems to be referring to here. Impossible and 'unscientific' as his claim may sound, given

of God is confined to one room of His vast mansion. He can learn by going from room to room in that mansion for each room is different from the other - that is true. If this was not so, God Himself would be barring Himself from His own creation.[55]

I have never denied communication[56] since I would be denying the action of God. But I do not uphold the common course used in spiritualism which is for the most part nothing more than a financial undertaking. And especially I do not uphold the possessiveness that expresses itself among those who practice it, when God possesses nothing. This possessiveness ofttimes is called 'earthbound'. This does not mean that spiritualism does not have a place in this world nor that we cannot communicate with those who have gone on, but it is the manner in which it is being handled.

Concerning your question on abortion - while it is not a good philosophy to preach this to the ignorance of man, since they are acting too extremely now in unnecessary form destruction, yet if we are to know the truth we must face it. To those of us who would not destroy for the

---

the fact that more and more of Adamski's teachings and claims are being corroborated by quantum research, systems science, world developments, and documentary evidence, some generosity in acknowledging that there are things about his mission that we do not yet understand seems more than justified. The esoteric explanation seems to be that bird migration from another planet to Earth, although not vice versa, may happen when the 'astral body' of another planet passes through planet Earth. See also Adamski's further comments in his letter of January 16, 1952, page 91.

55 Several sources indicate that Adamski himself originated from Venus and incarnated on Earth for his mission of informing humanity about the reality of the visitors from other planets. See: Aartsen (2015), *Priorities for a Planet in Transition*, p.149.

56 I.e. with 'higher' or extraterrestrial beings.

mere purpose of destruction, it is then true that all one destroys in such a procedure is the house within which the soul would have lived. While this is not murder in a true sense as we know it, it is denying that matter the elevation that it could have gotten through the soul manifesting in it. But actually to say that crime is committed, that would be awfully hard to admit. Pregnancy is but the natural sequence of an act that has been performed. It is the natural working of a universal law. In trees, flowers, vegetables, etc. we pull out the little ones that are too thick, out of place, and so on. In the human this is called 'abortion'. But here again division, created by man's opinion, has entered, for in the one case it is considered a crime, in the other it is a process of intelligent cultivation.

When man grows to the place of no longer worshipping or possessing forms but recognizes the occupant within the house, then will birth and death be no longer indulged in, but an eternal transmuting from state to state.

The Star of Bethlehem is composed of Sirius (Dog Star), the planet Venus and the North Star. In 1939 they were in a stage of conjunction for three months. Such a conjunction occurs only once in every 2,000 years. It takes 1,000 years for them to come together to that stage where to the naked eye they appear as one star, and another 1,000 years of retrogression.

You are right - the men from other planets who have come to Earth have the same type organs as Earth men and they breathe even as you and I. What I mean by 'allies' in another world is that there are groups or planets out there whose people are friendly to us - which doesn't mean to our nation itself but to the world of people.

About the space lab and airport in Australia, I will have more on that at a later date. One of my letters has been translated into Chilean for more information from there. A former commander of the Chilean Air Force spoke of this 2 years ago. Right now I haven't more information than I have already given you, but you might get more from your friend whose daughter-in-law is a native of Australia. On the other hand, a communication system is definitely going on, not only there but in this country as well. If you have seen the movie "The Day the Earth Stood Still" you can understand how they would understand us by monitoring us. They could come down here and speak our language or any Earth language - which they do. In fact, this picture[57] is more actuality than most people watching it realize. I knew more than 2 years ago of communication going on right here in southern California, and the very people who had that contact then - and it has been improved during this time - are the same group who made this picture. That is why the picture of the saucer is so realistic, as well as the space man with his philosophy. Also the picture surely did show up our military and the way fear propaganda is fed to the people by our newscasters. But let me repeat - the ship and the philosophy are real. I alone haven't this truth - there are many others who have this same information directly from the source who should know. The knowledge has been gained mostly from the established communication system, that is why the picture appears so real.

Yes, in time the public will be educated to where

57 I.e. motion picture.

the fright or the shock will not be so great as it would have been even a short time ago without the education. While there is not much being said about space ships in the newspapers any more, yet these visitors are moving through space observing our Earth closely and in great numbers right now. All these pictures dealing with space – with 7 more coming out shortly – are sponsored by the Air Force, I call it 'sponsored' since each picture has the signature of approval of the Air Force, which is for no other purpose than educational. TV also is doing its part in the educational program. So they really aren't doing as badly as it seems.[58] Remember, the public fears what it doesn't know. A United Press reporter told me just a short time ago that more than 95% of the people must be ready to accept the truth before it can be released. If it was less than this, and those fearing became panicky, they would make a mess in this world of ours. That is the reason not too much is coming out in these days – yet the educating program is going forward in picture form.[59] At the same time the files are being loaded

58 If this was indeed the US Air Force's attitude, and not Adamski's optimistic outlook, it soon changed when the Space Brothers' call for international cooperation and the abolishment of nuclear weapons met with growing interest from the public. Although the CIA-sponsored Robertson Panel concluded in 1953 that UFOs did not pose a threat to US national security, as UK researcher Robbie Graham documents, it still "suggested that the USAF begin a 'debunking' campaign employing the talents of psychiatrists, astronomers and celebrities, with the goal of demystifying UFO reports" and to "take immediate steps to strip the Unidentified Flying Objects of the special status they have been given..." (*Silver Screen Saucers* (2015), p.17) This change of heart was soon apparent even in the films that were released about the subject, with titles such as *Invaders from Mars*, *Earth vs the Flying Saucers*, *Invasion of the Saucer Men*, et cetera.
A letter to Gray Barker, editor of *The Saucerian*, of June 8, 1955, shows that by that time Adamski was aware that the media were used to influence people's perception: "Another lesson I learned: in Columbus, Ohio, I was told of a group, associates of the major one, who control information given to the people through the medium of newspapers, radio and TV..."

59 It seems as if Adamski was still unaware of the monumental task that lay in

with new reports and new photographs of space ships. In the next letter I may be able to give you more information in reference to things which are now shaping up.

In the meantime, I hope this gives you all the information you have been seeking plus the long awaited answers to your questions. Let yourself realize the blessed household within which you live, the vast universe holy wherever you might be. As it takes many types of furnishings to furnish the Earthly household, so it is in the universe, and every type of that furnishing is equally as holy as the room within which it is placed.

Very sincerely,

Prof. Geo Adamski

Professor George Adamski
Star Route,
Valley Center, California

---

store for him, and that would start with the newspaper report about his first contact in the desert that appeared on November 24, 1952, exactly one year after he wrote this letter. (See also pages 98-100.)

January 16, 1952

Dear Miss Martinelli,

Answering your letter of December 24 and speaking of visitors from other planets, you see, in the physical I have not contacted any of them, but since you have read Pioneers of Space you can see how I get my information about these people and their homelands.

On the other hand I will now tell you the latest information which came to me on Thanksgiving day. A very fine man stopped into our place Thanksgiving morning wanting something to eat - it was about an hour before noon and before the general public got started. I asked him if he had been up to the big observatory.

He said he was on his way up there.

I told him there was snow there and it might be a little chilly.

He replied, "Oh, I'm used to this kind of weather. Where I come from it is much colder. You see I am a marine engineer and my home is in Alaska. I work around the Aleutians."

So right there and then the thought struck me to ask him about flying saucers in that vicinity.

"Of course I have seen them, lots of them. Not only the flying saucers but the cigar-shaped type too. I have a two year old daughter who has seen more of them and has even been inside more of them than most people will ever see in all their lives. You see, they have been landing there," he said, but he wouldn't tell me the location of the base or whether there were more than one base.[60] "When they come in we usually drive out to see the ships and visit with the crews. My little girl is with me on most of these drives."

He told me the following straight through, as I asked different questions: all space craft are magnetically propelled; they vary in size all the way from 30 feet to 5 miles in length; no greater comfort or beauty could be found anywhere than is inside these ships - they are regular palaces. They are coming from planets Mars, Venus, Saturn and a system beyond ours known as system 359, the Wolf Star. The occupants of these ships speak the many languages of this world which they have learned through monitoring our radios and each ship has the insignia of its planet on it. These men from other worlds do not call planets by name as we call them; instead, they designate them as "worlds" in such and such orbit. The only ships that carry a definite insignia of its planet is Saturn - which is a circle with a line through it - while the others have the sun radiating as the center, encircled by planetary orbits with their home planet placed in its numerical orbit. That is: Venus is in the second orbit; Mars is orbit #4. The other system is indicated by a spiral insignia, he said, but didn't go into further detail.

The men average from 3 feet to 6½ feet in height[61],

60 See note 48 on page 72.

61 This is another instance where Adamski's statement predates the information of many other contactees from around the world.

are very handsome and awfully brilliant. As he said, we are a crude form of humanity alongside of them. When one is in their presence he feels like a babe before a great man of wisdom. The uniform of the tall men is a one-piece garment with an automatic button by which, when they press it the whole garment opens up and can be dropped off easily and quickly, if desired. The little men wear two-piece uniforms fastened together with the same type of fastener.

According to my informant, there seems to be no differentiation in rank among the space men - no higher nor lower, all are equal. And I believe they did have something to do, from what he told me, with the sudden change in our foreign policy whereby disarmament of the world suddenly became of paramount importance. So that must be their purpose in coming to earth so regularly - he wouldn't tell me much more than that on this particular subject.[62]

Then I showed him some of my pictures. He recognized them and wondered how I really got them as good as I did while they were out in space. The only thing lacking was markings and the insignia. So now I am trying awfully hard to get a picture with an insignia since each of these ships has one. If the ship presents itself at the right angle with the proper light on it, I believe I can get it. And when I do, that will break the wall of secrecy, at least to some extent if not all the way. I have a lot more to tell you in reference to space men and so forth, but I will leave that for another time.

Now I will continue with this information. I have a piece of magnetic turbine which fell out of the sky over

62 See also Adamski's reference to the military proving grounds on page 55.

Marion, Ohio in 1910. It looks quite crude although it went through terrific heat. I showed this to my visitor and he immediately recognized it as a part of a turbine out of one type of space ship. He said this particular turbine must have been made in an emergency during flight and it didn't hold up, so was thrown out. He explained that space ship crews can even cast parts while in flight and make any necessary parts to get home to their base where finely precision worked parts are put in to replace the damaged ones. This clarifies the Maury Island and Tacoma mystery of 1947 when metal and slack were dropped out of a space ship into the bay and onto the beach. Kenneth Arnold gives a full description of this incident in the January issue of Other Worlds.[63]

Besides the several types of space craft so far observed by earth men, there are also monstrous sized cruisers of space. None of these have yet landed on this earth. The cruisers, according to what I have been told, are of the size of a fair sized city. Now this might sound fantastic to us, but I have no reason to doubt this man for he did prove himself with his credentials to be what he said he was. As to the size, we should stop and think for a moment, the bigger the ships, the more comfortable with less awareness of movement for the passengers. Our planets are big aren't they? And they float like little balloons in space. Once the law is known, ships can do likewise. Speed? That too doesn't mean much even if it were one million miles per second, for we forget that the solar system within which we are living is travelling at 12 miles per second, while our earth within it is travelling 18 miles per minute,

63 *Other Worlds Science Stories*, a monthly magazine edited by Raymond Palmer from 1949 until 1953.

and do we feel it? No. All because of its size. In other words, most of our limitations are the result of improper knowledge, but once this is realized, then the incredible becomes credible.

So much for space cruisers.

Now in reference to birds migrating from other planets, this I have known for some time and only recently had it recalled to my attention, but I don't know where to refer you for definite data. They migrate from planets as they do from north to south and so forth; which proves also that the atmospheric conditions through space must be fair enough to be coped with, for the space men mentioned above don't wear any protection against our atmosphere either.

Of course the word 'teleportation' means transferring an object from place to place; at least that is what I think it represents. It is very odd how these mediums coin names for things because I am positively sure that the same type of names would not be given through a medium from another planet or from the universe. So they must be coined, if anything. Apport - I don't know why they should use this word for creation - it might mean presence also. That is the whole trouble with a lot of these good-meaning people. As though we didn't have enough mystery already about ourselves without creating any more of them! The simplicity seems to miss them all. The word 'transmigration' I think would be sufficient for most anything and its meaning is well understood by everybody. It doesn't make any difference whether you are transmigrated from this world unto another through the process of eliminating one body and going to someplace else, or whether you took the body with you and went somewhere else. It is still transmigrating. I don't believe that the world of people will be helped very much by coining new phrases for something that is very simple: for it is simplicity that will restore man to his original state.

Now a little bit more on abortion. As I stated before, I still stand on it, but here is the rub if it can be called a 'rub'. After all, man is a creator. What he fails to bring forth into creation, he fails to enjoy - naturally. So the only penalty one pays, if they take it that way, is that when they probably want a likeness of themselves in later years, they don't have it. But thousands of people don't really care; while others do. I for one, I might say, I don't really care whether I have a likeness of me left in this world or not, and especially when that likeness might have to face conditions in this world far worse than I have faced.[64] If I can't bring my likeness into something better, then I surely don't want to bring it into something worse. And since I am inclined for the better, I would be doing an injustice to my purpose if I brought into existence a being to experience conditions which are growing worse instead of better.

About my Chilean friend, I now have written four articles for Latin American papers and sent them to him as he requested. He is to contact the papers and get the articles published. This will take a little time since he must first translate them, but I did get a letter from him shortly before Christmas saying that his brother, who is a scientist for the Chilean government and prior to that, a

64 This seems to indicate Adamski was prepared for world conditions to worsen, despite the presence of the saucers, which he saw as a sign of a brighter future dawning.

commanding officer in their air force, is flying from Chile sometime in March to see me.[65] He must have much to say since he is flying here instead of writing: some of it must be quite important where writing would not be permissible.

I have not yet heard from the Holt Company about my book, but I have been told that the longer they hold it, the better the chances are that they are going to publish it.

Speaking of friends, yes a friend of yours, Rev. Nothdurft, has written me several letters and he has told me about your sharing your letters. Such a passing on of truth is good for all concerned. I sent Rev. Nothdurft a set of pictures for his private collection, but not for publicity of any kind at this time. I haven't heard from him since. That was almost three weeks ago. I shall get a set ready for you in the near future.

This will be about all for this time, will write more on the spiritual things next time. But just this reminder for now: there is one thing that a person should not forget – do not desire a distant place as the reward of goodness or good deeds.[66] Realize one thing: the Creator could not create anything lesser than Himself, therefore you are already complete and living in the holiest of holies. No greater holiness will ever be known. All you do is to realize this truth and then learn to live within it. For one to live in a palace, must know how to behave in a palace. The one who does not know, soon makes a mess out of the palace in which

65 It is interesting to note here that several South American governments, including Chile, have long openly acknowledged the presence of UFOs in their skies, according to the documentary *UFOs in South America* (2010).

66 Here Adamski seems to address Ms Martinelli's longing for a better world.

he lives. The universe is a divine household, or a palace, and we are all in it and can never walk out of it. That is how short the lesson of life really is, but a palace has many things within it and these cannot be learned of in a day. But eternity will reveal them all, so long as we know where we are.

Best wishes for your every good throughout the present new year.

Sincerely your friend,

Prof Geo Adamski

Professor George Adamski

From *LIFE* magazine, April 7, 1952 – 'Have we visitors from space?' (see next page):

This review has resulted from more than a year of sifting and weighing all reports of unexplained aerial phenomena—from the so-called flying saucers to the mysterious green fireballs so often sighted in the Southwest (*above*). This inquiry has included scrutiny of hundreds of reported sightings, interviews with eyewitnesses across the country and careful review of the facts with some of the world's ablest physicists, astronomers and experts on guided missiles. For the first time the Air Force (while in no way identifying itself with any particular conclusions) has opened its files for study.

Out of this exhaustive inquiry these propositions seem firmly shaped by the evidence:

1. Disks, cylinders and similar objects of geometrical form, luminous quality and solid nature for several years have been, and may be now, actually present in the atmosphere of the earth.
2. Globes of green fire also, of a brightness more intense than the full moon's, have frequently passed through the skies.
3. These objects cannot be explained by present science as natural phenomena—but solely as artificial devices, created and operated by a high intelligence.
4. Finally, no power plant known or projected on earth could account for the performance of these devices.

May 8, 1952

Dear Miss Martinelli:

It has been a long time since I have heard from you but tonight I have been answering some of my stacked up mail and a very strong thought came of you, so I am sending this on for what it is worth. For a while I thought I might be in your city for a lecture for the Universal Studies Forum. Rev. Nothdurft wrote that Mr. Gardner would like to have me come up, but thought Mr. Gardner intended writing directly to me. I waited for a letter from Mr. Gardner but when one did not come, I wrote to him. Now there is a tentative arrangement for an early fall series.

Just the other day I received a letter from Rev. Nothdurft which was most interesting and requires a special answer. I shall try to get that done within the next few days. He seems most interested in knowledge of the vastness of creation, which is most unusual, as I have found it, for a pastor of a Methodist Church. Orthodox churches of today need more men like him.

Well the flying saucers hit the headlines again, didn't they? By now you surely have read the official admission of them in LIFE[67], and I believe the 'skywatch' which becomes effective on the 17th of this month is associated with flying saucers for there is no indication of anything else. These visitors are coming in in greater numbers than ever before, yet published reports of their sightings are still scarce.

67 See the excerpt on page 94.

By the way, have you heard anything more about that space ship base in Australia? If so, I would like to have a little more information on it. Can you get me its approximate location?

I have just succeeded in getting a supply of my space craft pictures made up so now I can supply the demand that has been growing for them. Up till now I have only been able to get a few at a time. As you probably know, Rev. Nothdurft has a set, one of the very few that I sent out, and you requested notification when I had them for sale. To you I will send them at the same price that I was able to get the first few made - 35¢ each or the set of 12 for $4.20. However, with conditions as they are now, I must charge 75¢ each for these pictures to the general public, or the set of 12 for $7.20.

Now for the thought which was the original purpose of this letter. The question is, "Could a Creator create anything lesser than Himself?" and you might like to discuss it pro and con with your friends.

My answer is, "No".

Then what are the conclusions?

We are already as holy as holy can be, and we are already living in a heaven, or the spirit world, or the Infinite allness. There is really nothing more to attain, but what is to be done is for the human mind to realize this and to live that holiness without divisions by understanding the relative state of each form to all others, regardless what the form represents. For only through the alphabet can one spell out a whole word, and only through knowing the relative position of each form can one know the purpose of Divinity. The reason why living on earth is not heaven to most of us is because we have not learned how to live a heavenly life.

A person could move into the most magnificent palace ever built, but if he did not know how to live in a palace, that palace soon would become a common place. So we are already living in heaven, if that is the word chosen for the highest place of attainment, but we must learn how to live in it.

And since Divine Mind is not divided, divisions cannot prevail. Unified understanding of life as we find it is the only thing that will awaken us to the everlasting presence of manifestations without a beginning or ending, and the eternal plan of the Divine Mind. This means wisdom and understanding. Know thyself - and 'thyself' is all-inclusive - then thou shalt know all things. Knowing all things, mysteries cease to be and divisions are replaced by oneness.[68]

May the blessings of the Infinite be with you,

Prof Geo Adamski

Professor George Adamski
Star Route
Valley Center, California

68 "Knowing oneself" and one's "relative state" to all others in these last three paragraphs refers to learning to live in a "palace", i.e. in right relation with all of humanity, nature and the planet, which requires entirely new societal structures (see pages 22-25, 67 and 69). See also Aartsen (2015), *Priorities for a Planet in Transition*, chapter 3, 'Right human relations: An extraterrestrial show-and-tell'.

SIX MONTHS LATER . . .

First contact in the desert

On November 20, 1952 George Adamski, two coworkers and four friends drove to the Coxcomb Mountains in the California desert for a picnic, in hopes of spotting a flying saucer. Around noon, after first noticing a mothership, they saw a saucer landing in the nearby foothills. When Adamski went over for closer inspection he was met by one of the occupants. These historic events were described by Adamski in *Flying Saucers Have Landed* (1953), and meticulously documented by researcher Michel Zirger in *Authenticating the George Adamski Case* (2018).

George Hunt Williamson and his wife Betty were among the party who witnessed Adamski's encounter with a visitor from space. Shortly before the events transpired they took this photo, showing coworker Lucy McGinnis standing by the car on the left and Adamski sitting next to his 6-inch telescope.
Image: George Hunt Williamson (1953), *Other Tongues – Other Flesh*, p.96.

The Phoenix Gazette

ARIZONA'S PROGRESSIVE NEWSPAPER

nd-class matter under Act of Congress March 3, 1879. PHOENIX, ARIZONA, MONDAY, NOVEMBER 24, 1952 7 CENTS A COPY.

Flying Saucer 'Passenger' Declares A-Bomb Blasts Reason For Visits

By LEN WELCH

Fasten your safety belt, Buster, and take a firm grip on your chair for we are about to take off on a story to end all stories about flying saucers.

Woven into this incredible tale is what was reported to be probably the first person-to-person conversation with a man in a flying saucer, an explanation why flying saucers are flitting about our skies, a beautiful woman from another planet, and mysterious footprints in the desert sands.

Few questions about flying saucers are left unanswered by this story that has its beginning on a lonely spot on the California desert between Parker, Ariz., and Desert Center, Calif.

ITS PRINCIPALS are four Arizonans out to get a look at a flying saucer, a Valley Center, Calif., "professor," his secretary, and another woman, both from Valley Center.

The Arizonans figuring in the story are George Williamson, 25, Prescott, an employe of the supply division procurement section at the United States Veterans Administration Hospital at Fort Whipple; Mrs. Williamson, a medical technician employed in the laboratory at the hospital; Alfred C. Bailey, 38, of Winslow, for 12 years an employe of the Santa Fe Railway and now "braking" on passenger trains, and Mrs. Bailey.

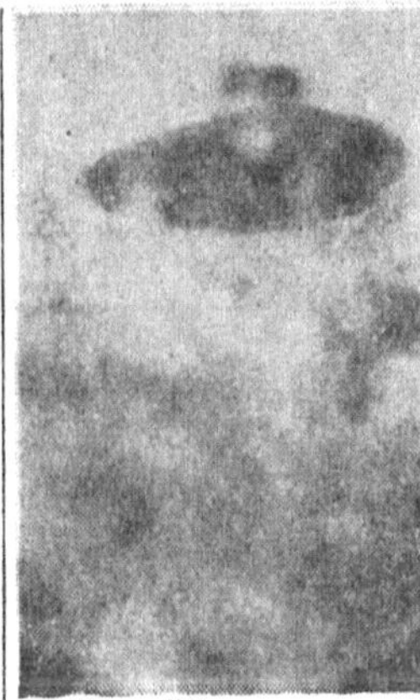

Is this a flying saucer or a freak cloud formation? Prof. George Adamski of Valley Center, Calif., gave the negative to Arizona friends, claiming he took the picture about 10 miles east of Desert Center, Calif., on the Desert Center-Parker Highway and later engaged in conversation with a man from space.

Williamson's interest in flying saucers was intensified by stories of saucers he uncovered among Indian legends while doing independent research among the Chippewas.

"I had corresponded with Prof. George Adamski, formerly of Palomar Observatory near San Diego, and learned that he had made pictures of flying saucers," Williamson said. "We (my wife and the Baileys) decided to go on a picnic lunch with Professor Adamski in the hope that we would see a flying saucer."

THE GROUP in addition to the Williamsons, the Baileys and Professor Adamski included Alice K. Wells and Lucy R. McGinnis, the latter the professor's secretary, both of Valley Center.

We will now proceed chronologically with the fantastic story of events that occurred as pieced together from stories of the Baileys and Williamsons. This is their version:

The party drove to a spot on the Desert Center-Parker Highway about 10 miles east of Desert Center. They parked

Turn to SAUCER on Page 16

Four days later, on November 24, 1952 the Arizonian *Phoenix Gazette* published a report based on the account of four of the eyewitnesses – George Hunt Williamson, his wife Betty, and Alfred and Betty Bailey.

Convinced that Adamski had given them an honest account of his experience, which they had witnessed from a distance, they recounted his telepathic exchange with the visitor as follows [quoted from the article]:

"Adamski: 'What is the purpose of your visits to earth?'

Visitor: Uses arms to indicate mushroom-shaped clouds

associated with atomic experiments convincing Adamski that such experiments are responsible for visits.

Adamski: 'Why are you concerned about these atomic experiments?'

Visitor: Indicates that radiation from explosions is causing his people some concern and fear that blasts will destroy everything. Adamski indicated he would like to prowl around in ship but visitor shook head indicating top secret stuff inside. (...)

Adamski: 'Did you come from the bigger ship?'

Visitor: 'Yes.'

Adamski: 'Are you from another planet?'

Visitor: Indicates yes but that he couldn't tell from where. (...)

Adamski then asked the visitor if he wouldn't permit his friends who were on their way to the spot to take some pictures but the visitor indicated 'no pictures of personalities right now.'

At this point some steps came out of the bottom of the saucer, the professor and the visitor shook hands and the visitor climbed into his machine and took off without a sound and vanished." [end of quote]

During a press conference in September 1955 Adamski explained what the effect of the report had been: "I will tell you honestly that I, too, would not have come out as much as I have, had it not been for that first contact and the four people who were with me … Williamson and the others (Mrs Williamson and Mr and Mrs Bailey), who went to Phoenix, Arizona, and gave the story to the *Gazette*. Once that came out I was on the spot completely and there was nothing else to do. And once you stick your neck out you might as well go ahead with it."[69]

---

69 Adamski (1955), *Many Mansions*, p.10.

APPENDIX

The Ageless Wisdom teaching and the Space Brothers

According to the Ageless Wisdom teachings the history of humanity on this planet goes back 18.5 million years, and the evolution of consciousness has not stopped with the emergence of the human kingdom from the mineral, vegetable, and animal kingdoms in nature.

Over time, out of the human kingdom has evolved the spiritual kingdom of the initiates and Masters of Wisdom, those members of the human kingdom who have gone ahead in evolution and have completed the planetary Path of Return in consciousness, or are in its final stages. Out of their midst, at the beginning of every new cosmic cycle or Age, a Teacher is sent into the world to reveal to humanity a further aspect of reality, of the Laws of Life, and our true spiritual Self, to guide and inspire humanity along the path of evolution to align our consciousness with the source of consciousness, and inaugurate a new dispensation.

The body of knowledge that is now referred to as the Ageless Wisdom teachings is nearly 100,000 years old, and was gradually put together by the enlightened Masters of the time.[1] The teachings were first reintroduced to the modern world in H.P. Blavatsky's books *Isis Unveiled* (1877) and *The Secret Doctrine* (1888), which presented this new perspective on history and human evolution. Her seminal teachings have been elaborated in many publications, notably those of Alice A. Bailey, the Agni Yoga series, and Benjamin Creme.

H.P. Blavatsky says that the term human "does not apply

1 Benjamin Creme, 'Questions and answers'. *Share International* magazine, Vol.33, No.7, September 2014, p.27.

merely to our terrestrial humanity, but to the mortals that inhabit any world, i.e., to those Intelligences that have reached the appropriate equilibrium between matter and spirit..."[2] As early as 1882 the Master Koot Hoomi referred to human and super-human beings on other worlds when He wrote about "all the intelligences that were, are or ever will be whether on our string of man-bearing planets or on any part or portion of our solar system".[3]

The Master Djwhal Khul (D.K.) states that "in all the [planetary] schemes, on some globe in the scheme, human beings, or self-conscious units, are to be found. Conditions of life, environment and form may differ, but the Human Hierarchy works in all schemes."[4]

Benjamin Creme added: "All Hierarchies of all the planets are in touch with each other, and everything that takes place in an extraterrestrial sense takes place under Law."[5]

Through his associates, the World Teacher said in April 1989: "Humanity has naively believed that they are the only ones in space. But there are others there, far advanced, who have always watched over us, teaching us not to kill, to respect others, and to learn to be happy and free."[6]

To this end, according to Creme, "The Space Brothers have on this planet various people, like [George] Adamski and others, who are used to bring the reality of the Space Brothers to the world..."[7]

Adapted from: *Our Elder Brothers Return – A History in Books (1875–Present)*, <www.biblioteca-ga.info>.

---

2 H.P. Blavatsky (1888), *The Secret Doctrine*, Vol.I, p.106.

3 A. Trevor Barker (1923), *The Mahatma Letters to A.P. Sinnet*, p.90.

4 Alice A. Bailey (1925), *A Treatise on Cosmic Fire*, pp.358-59.

5 Creme (2001), *The Great Approach – New Light and Life for Humanity*, p.129.

6 Brian James, 'Maitreya's view – The Age of Light'. *Share International* magazine, Vol.8, No.4, May 1989, p.5.

7 Creme (2010), *The Gathering of the Forces of Light*, p.36.

The Sea of Consciousness, feat. The Invisible Ocean

The Sea of Consciousness includes the integral text of George Adamski's lost debut *The Invisible Ocean*, two previously unpublished articles and a special clippings section documenting his time with the Royal Order of Tibet.

In three separate essays researcher Gerard Aartsen outlines how

- Adamski's philosophy is now being confirmed by 21st century science;
- historical facts expose the long-standing false allegations against Adamski; and
- science is edging ever closer to the recognition that the manifestation of life may not be limited to our carbon-based reality.

BGA Publications 2019, 118 pages.
ISBN: 978-90-9031695-6 (Paperback)

UFOs and the Pioneers of Oneness

NEW: Paperback edition

Pioneers of Oneness provides a groundbreaking synthesis of findings from systems science, quantum research, the wisdom teachings and contact experiences that strongly suggests, among other things, that Adamski's *Pioneers of Space* is based on actual (out-of-body) experiences.

The result is a riveting expose and a deep exploration into the Oneness of Life, the oneness of humanity, and our intricate connection with planet Earth and the Universe.

Pioneers of Oneness shows that a future filled with promise is indeed within our reach.

BGA Publications 2022, 276 pages.
ISBN: 978-90-830336-1-7 (Paperback edition)
ISBN: 978-90-830336-0-0 (Hardcover edition 2020)

RECOMMENDED READING

Priorities for a Planet in Transition

Packed with a mass of facts that have until now been ignored or only considered in isolation, *Priorities for a Planet in Transition* uncovers how the reports of the contactees are really about the crises facing humanity today.

This book gives us compelling insights into the causes of humanity's problems, as well as the solutions that the Space Brothers have been suggesting since the 1950s.

As it links together an abundance of sources, the author provides many answers and leaves us with only one question: 'How could we have missed this?'

BGA Publications 2015, 212 pages.
ISBN: 978-90-815495-4-7 (Paperback)

Here to Help: UFOs and the Space Brothers

Exploring the facts behind the myths, this book reframes the debate about the reality of the space visitors in view of the unprecedented changes engulfing the world today.

From historical records, documentary evidence and testimonies of early and contemporary contactees against the background of the Ageless Wisdom teaching, *Here to Help* uncovers the patient and sustained efforts of the space people to interact with people of Earth. In the process it reveals the true motive for their presence – to help humanity through this historical time of transition by sharing their wisdom and technology.

BGA Publications 2012, 200 pages.
ISBN: 978-90-815495-3-0 (Paperback)

www.ingramcontent.com/pod-product-compliance
Ingram Content Group UK Ltd.
Pitfield, Milton Keynes, MK11 3LW, UK
UKHW040031200726
13854UKWH00001B/475

9 789083 033624